WATER WARS

WATER WARS

Privatization, Pollution, and Profit

VANDANA SHIVA

North Atlantic Books
Berkeley, California

Published by
North Atlantic Books
Berkeley, California

Cover photo © maxim ibragimov/Shutterstock.com
Cover design by Jasmine Hromjak
Book design by Adept Content Solutions

Printed in the United States of America
Previously published in 2002 by South End Press.

Water Wars: Privatization, Pollution, and Profit is sponsored and published by the Society for the Study of Native Arts and Sciences (dba North Atlantic Books), an educational nonprofit based in Berkeley, California, that collaborates with partners to develop cross-cultural perspectives, nurture holistic views of art, science, the humanities, and healing, and seed personal and global transformation by publishing work on the relationship of body, spirit, and nature.

North Atlantic Books' publications are available through most bookstores. For further information, visit our website at www.northatlanticbooks.com or call 800-733-3000.

Library of Congress Cataloging-in-Publication Data

Names: Shiva, Vandana.
Title: Water wars : privatization, pollution, and profit / Vandana Shiva.
Description: Berkeley, California : North Atlantic Books, [2016] | Originally published: Cambridge, MA : South End Press, c2002. | Includes index.
Identifiers: LCCN 2015044148| ISBN 9781623170721 (trade paper) | ISBN 9781623170738 (e-book)
Subjects: LCSH: Water-supply--Economic aspects. | Water-supply--Political aspects. | Water resources development--Economic aspects. | Water rights. | Water--Pollution.
Classification: LCC TD345 .S525 2016 | DDC 333.91--dc23
LC record available at http://lccn.loc.gov/2015044148

1 2 3 4 5 6 7 8 United 21 20 19 18 17 16

Printed on recycled paper

CONTENTS

Water Wars

In 1995, Ismail Serageldin, vice president of the World Bank, made a much-quoted prediction about the future of war: "If the wars of this century were fought over oil, the wars of the next century will be fought over water." Many of the signs suggest that Serageldin is on target. Stories of water shortages in Israel, India, China, Bolivia, Canada, Mexico, Ghana, and the United States are making headlines in major newspapers, magazines, and academic journals.[1] On April 16, 2001, the *New York Times* featured a front-page story on water scarcity in Texas. Like Serageldin, the paper forecasted, "For Texas Now, Water, Not Oil, is Liquid Gold."[2]

While the *New York Times* and Serageldin are correct about water's importance in future conflicts, water wars are not a thing of the future. They already surround us, although they are not always easily recognizable as water wars. These wars are both paradigm wars—conflicts over how we perceive and experience water—and traditional wars, fought with guns and grenades. These clashes of water cultures are taking place in every society. Recently, when I was traveling to Rajasthan's capital, Jaipur, in western India, for a public hearing on drought and famine, I experienced the clash of these two cultures of water. On the train from Delhi to Jaipur, we were served bottled water, where Pepsi's water line Aquafma was the brand of choice. On the streets of Jaipur, there was another culture of water. At the peak of drought, small thatched huts called *Jal Mandirs* (water temples) were put up to give water from earthen water pots as a free gift to the thirsty, *Jal Mandirs* are a part of an ancient tradition of setting up *Piyaos*, free water stands in public areas. This was a clash between two cultures: a culture that sees water as sacred and treats its provision as a duty for the preservation of life and another that sees water as a commodity, and its ownership and trade as fundamental corporate rights. The culture of commodification is at war with diverse cultures of sharing, receiving, and giving water as a free gift. The nonsustainable, nonrenewable, and polluting plastic culture is at war with civilizations based on soil and mud and the cultures of renewal and rejuvenation. Imagine a billion Indians abandoning the practice of water giving at *Piyaos* and quenching their thirst from water in plastic bottles. How many mountains of plastic waste will it create? How much water will that dumped plastic destroy?

Paradigm wars over water are taking place in every society, East and West, North and South. In this sense, water wars are global wars, with diverse cultures and ecosystems, sharing the universal ethic of water as an ecological necessity, pitted against a corporate culture of privatization, greed, and enclosures of the water commons. On one side of these ecological contests and paradigm wars are millions of species and billions of people seeking enough water for sustenance. On the other side are a handful of global corporations, dominated by Suez Lyonnaise des Eaux, Vivendi Environment, and Bechtel and assisted by global institutions like the World Bank, the World Trade Organization (WTO), the International Monetary Fund (IMF), and G-7 governments.

Alongside these paradigm wars are actual wars over water between regions, within countries, and within communities. Whether it is in Punjab or in Palestine, political violence often arises from conflicts over scarce but vital water resources. In some conflicts, the role of water is explicit, as is the case with Syria and Turkey, or with Egypt and Ethiopia.[3]

But many political conflicts over resources are hidden or suppressed. Those who control power prefer to mask water wars as ethnic and religious conflicts. Such camouflaging is easy because regions along rivers are inhabited by pluralistic societies with diverse groups, languages, and practices. It is always possible to color water conflicts in such regions as conflicts among regions, religions, and ethnicities. In Punjab, an important component of conflicts that led to more than 15,000 deaths during the 1980s was an ongoing discord over the sharing of river waters. However, the conflict, which centered on development disagreements including strategies of the use and distribution of Punjab's rivers, was characterized as an issue of Sikh separatism. A water war was presented as a religious war. Such misrepresentations of water wars divert much-needed political energy from sustainable and just solutions to water sharing. Something similar has happened with the land and water conflicts between the Palestinians and Israelis. Conflicts over natural resources have been presented as primarily religious conflicts between Muslims and Jews.

Over the past two decades, I have witnessed conflicts over development and conflicts over natural resources mutate into communal conflicts, culminating in extremism and terrorism. My book *Violence of the Green Revolution* was an attempt to understand the ecology of terrorism. The lessons I have drawn from the growing but diverse expressions of fundamentalism and terrorism are the following:

1. Nondemocratic economic systems that centralize control over decision making and resources and displace people from productive employment and livelihoods create a culture of insecurity. Every policy decision is translated into the politics of "we" and "they." "We" have been unjustly treated, while "they" have gained privileges.

2. Destruction of resource rights and erosion of democratic control of natural resources, the economy, and means

of production undermine cultural identity. With identity no longer coming from the positive experience of being a farmer, a craftsperson, a teacher, or a nurse, culture is reduced to a negative shell where one identity is in competition with the "other" over scarce resources that define economic and political power.

3. Centralized economic systems also erode the democratic base of politics. In a democracy, the economic agenda is the political agenda. When the former is hijacked by the World Bank, the IMF, or the WTO, democracy is decimated. The only cards left in the hands of politicians eager to garner votes are those of race, religion, and ethnicity, which subsequently give rise to fundamentalism. And fundamentalism effectively fills the vacuum left by a decaying democracy. Economic globalization is fueling economic insecurity, eroding cultural diversity and identity, and assaulting the political freedoms of citizens. It is providing fertile ground for the cultivation of fundamentalism and terrorism. Instead of integrating people, corporate globalization is tearing apart communities.

The survival of people and democracy is contingent on a response to the double fascism of globalization—the economic fascism that destroys people's rights to resources and the fundamentalist fascism that feeds on people's displacement, dispossession, economic insecurities, and fears. On September 11, 2001, the tragic terrorist attacks on the World Trade Center and at the Pentagon unleashed a "war against terrorism" promulgated by the US government under George W. Bush. Despite the rhetoric, this war will not contain terrorism because it fails to address the roots of terrorism—economic insecurity, cultural subordination, and ecological dispossession. The new war is in fact creating a chain reaction of violence and spreading the virus of hate. And the magnitude of the damage to the earth caused by "smart" bombs and carpet bombing remains to be seen.

The Ecology of Peace

On September 18, 2001, I joined millions of people around the world to observe two minutes of silence in remembrance of the thousands of people who lost their lives in the September 11 assault on the World Trade Center and the Pentagon. But I also thought of the millions who are victims of other terrorist-actions and other forms of violence. And I renewed my commitment to resist violence in all its forms. That morning, I was with three women, Laxmi, Raibari, and Suranam, in Jhodia Sahi village in Orissa. Laxmi's husband, Ghabi Jhodia, was among the 20 tribals who have recently died of starvation. In the same village, Subarna Jhodia had also died. Later on that day, we met Singari in Bilamal village who had lost her husband Sadha, elder son Surat, younger son Paila, and daughter-in-law Sulami. World Bank-imposed policies had weakened the food economy and left these villagers vulnerable to famine.

Giant mining companies such as Hydro of Norway, Alcan of Canada, and Indico and Balco/Sterlite of India have joined the pulp industry to unleash a new wave of terror. They have their sights set on the bauxite resting in the majestic hills of Kashipur. Bauxite is used for aluminum, and aluminum is used for Coca Cola cans, rapidly displacing India's water culture, and for fighter planes, like those that are carpet-bombing Afghanistan as I write this. In 1993, we stopped the ecological terrorism of the mining industry in my home, the Doon Valley. The Indian Supreme Court closed the mines, ruling that commerce that threatens life must be stopped. But our ecological victories of the 1980s were undone with the environmental deregulation accompanying globalization policies. The aluminum companies want the homelands of the Kashipur tribals, and a major battle has ensued between residents and corporations.

This forced apportion of resources from people is a form of terrorism—corporate terrorism. I had gone to offer solidarity to victims of this corporate terrorism, which was not only threatening to rob 200 villages of their survival base but had already robbed many of their residents of their lives when they were shot and killed by the police on December 16, 2000. The 50 million Indian tribals who have been flooded out of their homes by dams over the past four decades were also victims of terrorism—they have faced the terror of technology and destructive development. The thirty thousand people who died in the Orissa Supercyclone and the millions who will die when flood and drought and cyclones become more severe are also enduring terrorism by climate change and fossil fuel pollution.

Destruction of water resources and of forest catchments and aquifers is a form of terrorism. Denying poor people access to water by privatizing water distribution or polluting wells and rivers is also terrorism. In the ecological context of water wars, terrorists are not just those hiding in the caves of Afghanistan. Some are hiding in corporate boardrooms and behind the free trade rules of the WTO, North American Free Trade Agreement (NAFTA), and Free Trade Area of the Americas (FTAA). They are hiding behind the privatization conditionalities of the IMF and World Bank. By refusing to sign the Kyoto protocol, President Bush is committing an act of ecological terrorism on numerous communities who may very well be wiped off the earth by global warming. In Seattle, the WTO was dubbed the "World Terrorist Organization" by protestors because its rules are denying millions the right to a sustainable livelihood.

Greed and appropriation of other people's share of the planet's precious resources are at the root of conflicts, and the root of terrorism. When President Bush and Prime Minister Tony Blair announced that the goal of the global war on terrorism is the defense of the American and European "way of life," they are declaring a war against the planet—its oil, its water, its biodiversity. A way of life for the 20 percent of the earth's people who use 80 percent of the planet's resources will dispossess 80 percent of its people of their just share of resources and eventually destroy the planet. We cannot survive as a species if greed is privileged and protected and the economics of the greedy set the rules for how we live and die.

The ecology of terror shows us the path to peace. Peace lies in nourishing ecological and economic democracy and nurturing diversity. Democracy is not merely an electoral ritual but the power of people to shape their destiny, determine how their natural resources are owned and utilized, how their thirst is quenched, how their food is produced and distributed, and what health and education systems they have.

As we remember the victims of September 11, 2001 in the United States, let us also strengthen our solidarity with the millions of invisible victims of other forms of terrorism and violence that threaten the very possibility of our future on this planet. We can turn this tragic and brutal historical moment into building cultures of peace. Creating peace requires us to resolve water wars, wars over food, wars over biodiversity, and wars over the atmosphere. As Gandhi once said, "The earth has enough for the needs of all, but not the greed of a few." The water cycle connects us all, and from water we can learn the path of peace and the way of freedom. We can learn how to transcend water wars created by greed, waste, and injustice, which create scarcity in our water abundant planet. We can work with the water cycle to reclaim water abundance. We can work together to create water democracies. And if we build democracy, we will build peace.

1. For articles on the water crisis featured in major publications in 2001, see Sandra L. Postel and Aaron T. Wolf, "Dehydrating Conflict," *Foreign Policy*, September/November 2001, p. 60; "Crazed by Thirst: Canadians are in Lather Over Water Exports," *The Economist*, September 15, 2001, p. 34; Nicholas George, "Billions Face Threat of Water Shortage," *Financial Times*, August 14, 2001, p. 6; "Water in China: In Deep," *The Economist*, August 18, 2001; "Low Water," *Financial Times*, August 14, 2001, p. 12.
2. Jim Yardley, "For Texas Now, Water, Not Oil, Is Liquid Gold," *New York Times*, April 16, 2001, Al.
3. See chapter 3 of this book for a more detailed discussion of the water conflicts in these countries.

INTRODUCTION TO
THE SECOND EDITION

Water *is* life. Water is the thread that interconnects life itself—the forests, the soil, the atmosphere, plants, animals, and human beings. Water connects us all. We are primarily water, and we are all sustained by it. Water is a commons.

Because cooperation and self-governance are vital to protecting water as a commons, water creates conditions of peace. When water disappears, or competition over scarce water resources grows, conflicts and wars are the result. Many conflicts and wars of our times are Water Wars—but as a matter of convenience, they are given cultural and religious labels and pushed into becoming religious conflicts. Water Democracy is at the heart of sustainability, justice, and peace.

In the many years since I first wrote *Water Wars*, the lessons from water have become more evident, more sharp, more intense, more vivid. The processes leading to destruction, pollution, and privatization of water have intensified and accelerated, and so have the movements to protect water and defend its existence as a commons.

Every time nonsustainable human activity disrupts the earth's potential for renewing life's processes, we disrupt the water cycle and water pathways. Seen from the perspective of water, every violation of the water cycle is an act of war and violence against the

Earth and life itself. Every time nonsustainable and unjust paradigms of water or land use enclose the water commons, crippling the capacity of ecosystems to sustain themselves and undermining the sustenance of communities and entire societies, it is a Water War. This triggers new conflicts and violence.

In the last decade and a half since *Water Wars* was first published, the two big movements for water democracy in which I have been engaged have been the women's movement against Coca Cola in Plachimada and against the World Bank–driven water privatization of Delhi's water. The Coca Cola plant in Plachimada has been shut down, and Suez—the multinational corporation attempting to privatize Delhi's water—was not able to pirate Delhi's water supply. Most privatization of water projects pushed by World Bank and the International Monetary Fund (IMF) have been stopped wherever and whenever movements for water democracy have arisen to defend the rights of water, the rights of communities to their water commons, and the right to life.

Unnatural Disasters as Water Wars: Climate Change, Maldevelopment, and Disasters

Maldevelopment and climate chaos have combined to aggravate the impact of disasters. I prefer to talk of climate "chaos," not climate change, for two reasons: "change" suggests predictability, and change in itself is not a bad thing. What we are experiencing is climate chaos—there is nothing predictable anymore about when the rains will come, where they will come, how intensely they will come. The high-altitude desert of Ladakh which gets no rain has had intense floods in 2007, 2010, and 2015. Uttarakhand—my home region—had such extreme rain on June 16 and 17, 2013 that settlements, homes, schools, roads, bridges, and thousands of lives were washed away.

Uttarakhand is the source of the sacred Ganga and its tributaries. The sources of the Ganga—the lifeline of India—were designated sacred sites in order to protect the Ganga Himalaya, and hence India. The *yatra* to the four pilgrimage centers of Gangotri, Yamunotri, Kedarnath, and Badrinath—the *Char Dham*—was meant to both connect us culturally and spiritually to these "Bhu tirths" (sacred sites of the Earth) and to connect us ecologically to the sources of life, the sources of our rivers, the sources of a civilization's water.

June 2013 was peak pilgrimage season in the *Char Dham*, including Kedarnath—which was washed away in its entirety—the lone ancient temple standing in defiance amidst the rubble left behind by the flood.

While the melting of snow in the Arctic has attracted a lot of attention, the snow melt in the Himalaya—the Third Pole—has been largely ignored, even though the Himalaya and Himalayan river systems support half of humanity. The melting of the Himalayan glaciers first creates glacial lakes that lead to major floods when they burst. Over time, as glaciers recede, perennial snow fed rivers become seasonal rivers, affecting agriculture, livelihoods, and society at large.

In Ladakh, a high-altitude alpine desert, glaciers, which are the only lifeline for the villages, are disappearing. Instead, Ladakh had heavy rainfall and floods in 2007, and then in 2010, the might of the water erased entire villages from the fragile landscape.

The retreat of the Himalayan glaciers is already taking place. The Gangotri glacier, from which the sacred Ganges originates, is retreating at the average rate of 27 meters per year. The Satopanth glacier is retreating at 26.9 meters per year and Dokriani at 15.7 meters per year. Chorabari—which feeds into the river Mandakini—is retreating at 10.2 meters annually. This snow melt has created the Chorabari Lake (3,960 meters above sea level), also known as Gandhi.

Sarovar is located about 4 kilometers upstream of Kedarnath, which is approximately 400 meters long and 200 meters wide, having a depth of 15–20 meters.

On June 16 and 17, 2013, for more than 72 hours, intense and unprecedented rain poured down on the Himalayan range. The Chorabari Lake burst, leading to its complete draining within 5–10 minutes, as reported by the watch and ward staff of the Wadia Institute of Himalayan Geology (WIHG), who were present at the WIHG camp at Chorabari Glacier at the time. Usually floods come at the end of a heavy monsoon. In 2013, they came with the first rain. The monsoon came early, and the rainfall was far greater than usual.

This is climate chaos.

These are climate disasters and yet, just before the Copenhagen Climate Conference, the Government of India issued a report saying there was no impact on India's glaciers. The Kedarnath tragedy shows how heavy the cost of this denial is. We need to recognize that our glaciers are threatened, and that melting glaciers will lead to more disasters, which will be greater in their might than we have seen. Disaster preparedness is the duty of government. But disaster preparedness needs honest and robust ecological science, and honest and robust participatory democracy.

The heavy rainfall together with melting of snow in the surrounding Chorabari Lake washed away the banks of the Mandakini River, causing massive devastation to Kedarnath valley and surrounding areas. The flood damaged the banks of River Mandakini for 18km between Kedarnath and Sonprayag, completely washing away the towns of Gaurikund (1,990 meters above sea level), Rambara (2,740 meters above sea level), and Kedarnath (3,546 meters above sea level). The roads and footpath between Gaurikund and Kedarnath were also damaged, leaving no way for the injured and stranded to be rescued for days afterward. According to local people, more than 20,000 people died in the Uttarakhand disaster of 2013.

June 2013 was more than a natural disaster. Climate change combined with maldevelopment to create the unprecedented disaster. The intense rain in a short period was typical of the extreme events climate change is bringing. And its impact was aggravated by the hydropower projects under construction in river valleys in Uttarakhand. The 99 Mega Watt power project at Singoli—Bhatwari, near Augustmuni, being constructed by Larson & Toubro, was a major cause of the devastation in Mandakini Valley in 2013. It is one of the 12 hydroelectric projects coming up on the Mandakini.

The ecological damage caused by maldevelopment has reduced the capacity of the mountain ecosystem to deal with heavy rain. Climate havoc adds to this vulnerability. The worst landslides occurred where tunnels were being built for these hydro-electric projects. Blasting the fragile mountains with dynamite, recklessly, for the construction of dams and tunnels has triggered thousands of landslides. When the first rain comes, these landslides fill the riverbed with rubble. There is no space for the water to flow. We are literally stealing the ecological space from our rivers and when they have no space to flow, they *will* overflow, cut banks and cause flooding.

The debris from the tunnels was dumped in the riverbed as well. When the floods carried the debris in the flood waters, the riverbed rose by 20 to 30 feet at places, the towering river waters washing away homes, villages, schools, roads, and bridges, killing the pilgrims as well as local people, who were all trapped. This destruction was aggravated by ignorance and greed.

We need to learn, once again, to have reverence for our sacred mountains and rivers. We need to be informed by the latest of ecological sciences, not by an obsolete "development" model, which is nothing more than an exploitation model that has led to tragic disasters, like the floods in Uttarakhand. The disaster is clearly manmade, climate change having been caused by man and made worse by man's addiction to maldevelopment. Politicians, decision makers, and businesses need to take responsibility for the disaster their actions, policies, and greed have caused.

Today, driven by greed and corruption, the government has become ignorant of the Indian culture of the sacred and the ecological fragility of the Himalayas. The sacred sets limits; ecological fragility sets limits. Today these limits are being violated, as rivers are dammed and diverted for electricity, and the pilgrimage to the Char Dhams is being turned into crass, consumerist mass tourism.

In 1916, in his book *Garhwal, Ancient and Modern*, Rai Patiram Bahadur wrote, "We may say that there is no country in the world of the dimension of Garhwal, which has so many rivers as a traveler will find in this land. The district has 60 rivers of different size; besides these, there are rivulets, rills, springs and fountains in hundreds, showing that nature has been especially bountiful to this land in the matter of its water supply" (quoted in Semwal, 21).

Five hundred dams are planned in our region on the Ganga system. Swami Gyanswarup Sanand (Formerly Dr. G. D. Agrawal) has been repeatedly going on fast to save the Ganga. His efforts forced the central government to declare the area from Uttarkashi to Gaumukh an ecologically fragile zone. The present Chief Minister has been blocking it in the name of "development." I hope that the disaster of 2013 will make him realize the value of protecting the Ganga Himalaya as an ecologically fragile zone. And it is not just the stretch between Uttarkashi and Gaumukh. We need to protect the entire catchment of the Ganga system as a cultural heritage and ecologically fragile, sensitive ecosystem.

Forty years ago, I joined the Chipko movement as a volunteer. The women-led Chipko movement started after the 1972 Alaknanda disaster, which was caused by logging in the Alaknanda valley. Women connected the deforestation to landslides and flooding. As they pointed out, the primary products of the forest were not timber and revenue, but soil and water. Forests left standing to protect the fragile Himalayan slopes provide more to the economy than when they are extracted as dead timber.

It took the 1978 Uttarkashi disaster for the Government to recognize that the women were right, when the government had to spend much more on flood relief than the revenues they were getting from timber extraction. In 1981, in response to the

Chipko movement, logging was banned in the Garhwal Himalaya above 1,000 meters. Today government policy recognizes that forestry in the fragile Himalaya has to be conservation forestry, which maximizes the ecological services of the forest with protecting—not extracting—forestry. In 1983, the Supreme Court of India ordered a stop to limestone mining in Doon Valley, recognizing that the limestone that was left in the mountains contributed more to the economy than the limestone that was extracted through mining.

The 2013 disaster should wake us up to the social, ecological, and economic costs of destructive policies which have devastated our fragile and beautiful mountain ecosystems. The Himalaya are the youngest mountain system in the world. They cannot bear the violence of deforestation and dam building. They need gentleness and respect. Chipko shook our policy makers out of their slumber, which had allowed them to think of forests as timber mines, and woke them to the ecological functions of the forests in the catchments of our rivers. The 2013 disaster should shake them out of their current slumber, which is allowing them to see rivers as nothing more than 20,000 megawatts of hydropower. It should help them to realize that when respected, our rivers are rivers of life, and when violated, they can become rivers of death.

Industrial Agriculture, the Climate, and Water Crisis

In the previous edition of *Water Wars,* I wrote about how the climate crisis is not just an issue of increasing temperatures, but also of climate extremes. Climate extremes are about too little or too much water. As I cited from an old Oriya expression:
  "Jala bhule, shrustinasa, jala bihune srustinasa."
  Too much or too little water destroys creation.
  Intensification of droughts, floods and cyclones is one of the predictable impacts of climate change and climate instability. In the final analysis, India's food security rests on the monsoon.

Monsoon failure and widespread drought imply a deepening of the already severe food crisis, which was triggered by trade liberalization policies and has made India the capital of hunger (Navdanya, "Why is every 4th Indian Hungry," 2009). It also implies the deepening of the water crisis that compelled me to write *Water Wars*.

The monsoons recharge the groundwater and surface water systems. When there is a drought, there will be reduced recharge. Since 1966, as a consequence of the introduction of the Green Revolution model of water-intensive chemical farming adopted under World Bank and US pressure, India has over-exploited her groundwater, creating a water famine. I had addressed this crisis in my 1984 book *The Violence of the Green Revolution*. Chemical monocultures of the Green Revolution use ten times more water than the biodiverse ecological farming systems the Green Revolution replaced.

In the 1970s, the World Bank gave massive loans to India to promote groundwater mining. It forced states like Maharashtra to stop growing water-prudent millets, like jowar—which needs 300 mm of water—and shift to water-guzzling crops like sugarcane—which needs 2500 mm of water. In a region with 600 millimeters of rainfall and 10 percent groundwater recharge, this is a recipe for water famine (see Navdanya's "Financing the Water Crisis").

A new study led by Matthew Rodell of NASA's Goddard Space Flight Center in Maryland and published in *Nature* has shown that water levels in North India have fallen by 1.6 inches (4 centimeters) per year between August 2002 and August 2008. More than 26 cubic miles (109 cubic km) of groundwater have disappeared from aquifers between 2002 and 2008. Most of this groundwater has been extracted for chemical Green Revolution –style farming.

Soil is the biggest water reservoir we have. When we return organic matter to the soil, we increase the capacity of the soil to hold moisture. Soil moisture is the most reliable drought and climate insurance. As Andre Leu, President of the International Federation of Organic Agriculture Movements, has shown:

- Volume of Water Retained /ha (to 30 cm) in relation to soil organic matter (OM).

- 0.5% OM = 80,000 liters (common conventional level)

- 1 % OM = 160,000 liters (common conventional level)

- 2 % OM = 320,000 liters

- 3 % OM = 480,000 liters

- 4 % OM = 640,000 liters

- 5 % OM = 800,000 liters

http://www.organicandclimate.org/Bileadmin/documents _organicresearch/rtoacc/events/11-IFOAM-rtoacc-20130712.pdf

Industrial agriculture destroys the water-holding capacity of soil. Hence, it requires more external inputs of irrigation. While contributing to climate change, it also makes agriculture more vulnerable.

Not only has chemical agriculture mined and wasted ground-water, it has also mined soil fertility and contributed, in great part, to climate change. Chemical fertilizers destroy the living processes of the soil and make soils more vulnerable to drought. Chemical fertilizers also produce nitrogen oxygen, a greenhouse gas that is 300 times more potent than carbon dioxide.

The solution for the climate crisis, the food crisis, or the water crisis under which the world is reeling are the same—biodiversity-based organic farming systems. Biodiverse ecological farms address the climate crisis by reducing emissions of greenhouse gases such as nitrogen oxide and absorbing carbon dioxide in plants and in the soil. Biodiversity and soils are the most effective carbon sinks. They also adapt to climate change and drought by increasing organic matter in soil, which increases the moisture-holding capacity of soil, and hence drought-proofs our agriculture.

Biodiverse organic farms increase food security by increasing the resilience and reducing the climate vulnerability of farming systems. They also enhance food security because they have higher production of food and nutrition per acre than Green Revolution monocultures, which measure the yield of one commodity, not the total food output, nor the nutritional quality of food.

Biodiverse organic systems also address the water crisis. Firstly, production based on water-prudent crops like millets reduces water demand. Secondly, organic systems use ten times less water than chemical systems. Thirdly, by transforming the soil into a water reservoir through increasing its organic matter content, biodiverse organic systems reduce irrigation dependence and help conserve water in agriculture. Thus, maximizing biodiversity and organic matter production simultaneously increases climate resilience, food security, and water security.

The 1965–1966 drought in India was used to push the Green Revolution, which has increased vulnerability to drought. The 2009 and 2015 droughts, and the climate crisis, are similarly being used to push the second Green Revolution with GMO seeds and patents on seeds. This will deepen Indian agriculture's vulnerability to drought.

The severe and frequent droughts in India and other countries is an opportunity to put water conservation in place, and through agriculture, at the center of agricultural policy. Instead of selling costly seeds and chemicals to produce nutritionally empty commodities, we must grow nutritional food, using less water while increasing the soil's capacity to hold water for times of scarcity. It is vital that Governments not allow themselves to be used by corporations as a marketer of GM seeds and Roundup in the midst of a water and climate emergency, to not allow the establishment of soil and climate data monopolies under the garb of climate-smart agriculture.

Having contributed to the creation of the crisis, corporations who have profited from chemical industrial agriculture are attempting to turn the climate crisis into an opportunity to control biopiracy-based climate-resilient seeds and climate data, while attempting to criminalize genuinely climate-resilient biodiverse organic agriculture. Monsanto now owns the world's biggest climate data and soil data corporations. Armed with proprietary big data, Monsanto is ready to profit from a crisis once more. The worse it gets, the better it is for Monsanto. Mitigating the crisis would not be profitable to climate deniers like Monsanto.

The Gates Foundation and the fertilizer and biotech industry—the Exxons of agriculture—joined hands at the Paris climate summit to push the false solution of climate-smart agriculture. The Gates Foundation, along with the other biotech evangelists of our times, have it completely wrong—climate-smart agriculture and "One Agriculture," packaged in a PR bubble, will starve the world and worsen the refugee crisis, which is already spiraling out of control. The Gates Foundation, pretending to feed the world, is proselytizing the very model of agriculture that has contributed to half of the climate problem as a solution.

One Agriculture, for the profit of one company, is hardly a mitigation strategy for climate chaos. Agroecology is already helping check climate change, by converting fossilized carbon to Green Carbon. Every seed is an embodiment of millennia of nature's evolution and centuries of farmers' breeding. It is the distilled expression of the intelligence of the earth and intelligence of farming communities. Farmers have bred seeds for diversity, resilience, taste, nutrition, health, and adaption to local agro-ecosystems. In times of climate change we need the biodiversity of farmers' varieties to adapt and evolve. Climate extremes are being experienced through more frequent and intense cyclones which bring salt water to the land. For resilience to cyclones we need salt-tolerant varieties, and we need them in the commons.

The delusional corporations have not "invented" climate-resilient traits in seeds. They have simply pirated the traits from farmers' varieties. Fifteen hundred patents on climate-resilient crops have been taken out by big biotech. Navdanya/Research

Foundation for Science, Technology, and Ecology, have published the list in the report "Biopiracy of Climate Resilient Crops: Gene Giants Steal Farmers Innovation." With these very broad patents, corporations like Monsanto can prevent access to climate-resilient seeds in the aftermath of climate disasters through patents—which grant an exclusive right to produce, distribute, sell the patented product. Climate-resilient traits are not created through genetic engineering, they are pirated from seeds farmers have evolved over generations. For thousands of years, farmers, especially women, have evolved and bred seed freely in partnership with each other and with nature, to further increase the diversity of that which nature has given us and adapt it to the needs of different cultures.

Biodiversity and cultural diversity have mutually shaped one another over time. Along coastal areas, farmers have evolved flood-tolerant and salt-tolerant varieties of rice—such as *Bhundi, Kalambank, Lunabakada, Sankarchin, Nalidhulia, Ravana, Seulapuni*, and *Dhosarakhuda*. After the Orissa Supercyclone, Navdanya could distribute two trucks of salt-tolerant rices to farmers because we had conserved them, as a commons, in our community seed bank in Orissa.

What needs to be done is clear. In the case of climate change, the key strategy should be a reduction of emissions and strategies for adaptation. We must move away from industrial, chemical-intensive agriculture and away from a centralized, global commodity-based food system that exacerbates emissions. Biodiversity conservation will be central to adaptation. In place of the biodiversity destroying industrial monocultures, including those based on GMO seeds, we need a shift to agroecological practices that conserve biodiversity and ensure biosafety.

We cannot depend on a mechanistic mind and its unscientific denial of the sciences of the interconnected nature of living systems and ecosystems to get us out of the crisis it has created. As Einstein said, "We cannot solve a problem with the same mindset that created it." Centralized, monoculture-based, fossil fuel-intensive petrochemical systems, including that of GMO agriculture, are not flexible. They cannot adapt and evolve. GMO technology has failed across the world. We need flexibility, resilience, and adaptation to a changed reality. This resilience comes from diversity. This diversity of knowledge, economics, and politics is what I call Earth Democracy.

Desertification, Displacement, and Conflicts: The Emerging Face of Water Wars

When we make a transition from chemical agriculture to ecological, organic agriculture, we conserve water by making soil a water reservoir. This reduces demands for irrigation, making more water available to sustain ecosystems and societies. Industrial globalized agriculture is using up more than 75 percent of the global water supply for irrigation. Through the combination of water exploitation and climate change induced extreme droughts, it also contributes to soil degradation, desertification and displacement of people from their land when the land can no longer support life.

In Terra Viva, a Manifesto we released at the Expo in Milan on 2nd May 2015, we have traced how the refugee crisis, which has displaced millions from Syria, and is now threatening to destabilize Europe, can be traced to the soil and water crisis, and of course the resulting wars and conflicts.

Across the world we witness new violent conflicts emerging as ecological consequences of the predatory economic model. According to the United Nations Convention to Combat Desertification (UNCCD), 40 percent of the intrastate conflicts over a 60-year period were associated with land and natural resources.

Moreover, 80 percent of the major armed conflicts in 2007 occurred in vulnerable dry ecosystems. Whether it is the rise in violence in Punjab in 1984, or in Syria and Nigeria today, the conflicts originate in the destruction of soil and water, and the inability of land to sustain lives and livelihoods any more. Unfortunately, however, the conflicts are not seen in their ecological contexts and are relegated instead to religious motives, with more violence and militarization offered as solutions.

The exposure of more and more people to water scarcity and hunger opens the door to the failure of fragile states and regional conflicts. In sub-Saharan Africa, the combined challenge of an increased population, demands on natural resources, and the effects of climate change (particularly drought) on food and water supplies are likely to lead to tensions, which could result in conflict.

The convergence of financial, food, climate, and energy crises impacts soils and peoples in many ways. Coupled with wars, these translate into waves of internally and externally displaced persons. Uprooted people are vulnerable to other exploitations, and the soils they once knew and defended are open for appropriation, despoliation, and general harm. In rural areas where people depend on scarce productive land resources, land degradation is a driver of forced migration. An estimated 42 percent of households intensify their seasonal mobility in the event of poor harvests, while 17 percent migrate when there is crop destruction, and 13 percent leave in the case of strong climatic events such as extreme droughts. By 2050, 200 million people may be permanently displaced environmental migrants.

Syria is part of the fertile crescent where agriculture evolved. Many of the crops that are now the staples of humanity are gifts from this region. For thousands of years, farmers sustained the soil and water. A few decades of the spread of uniform seeds bred for chemicals has drained the soils and groundwater. The extended drought between 2006 and 2009 triggered displacement; the refugee crisis; and the consequent conflicts, wars, and extremism that have now engulfed our world. Before the Syrian uprising of 2011, 60 percent of Syria's land experienced the most severe and prolonged drought, causing crop failures in the land where agriculture began and has endured for 12,000 years.

The impact of the drought was aggravated by nonsustainable use of land and water through the promotion of nonsustainable, chemical-intensive industrial agriculture. More than 80 percent of crops failed and more than 75 percent of livestock died, wiping out livelihoods and forcing a mass migration of more than a million farmers and herders to cities unable to handle the influx, resulting in social instability and the country's civil war.

Syrian Refugees—A snapshot of the crisis in the Middle East and Europe, by the Migration Policy Centre European University Institute, Florence, 2013.

http://www.nytimes.com/2015/03/03/science/earth/study -links-syria-conflict-to-drought-caused- by-climate-change. html?_r=0

http://www.historicalclimatology.com/blog/is-climate-change -behind-the-syrian-civil-war http://www.voanews.com/con- tent/drought-called-factor-in-syria-uprising/1733068.html

http://climateandsecurity.org/2012/02/29/syria-climate -change-drought-and-social-unrest/

An estimated 9 million Syrians have fled their homes since the outbreak of civil war in March 2011, taking refuge in neighboring countries or within Syria itself. According to the United Nations High Commissioner for Refugees (UNHCR), over three million people have fled to Syria's immediate neighbors: Turkey, Lebanon, Jordan, and Iraq. Six and a half million people are internally displaced within Syria. A million refugees have arrived in Europe by sea, and 34,000 over land. Many lost their lives, including little Aylan Kurdi, whose dead body washed ashore on a beach and reminds us to change our ways.

http://syrianrefugees.eu

http://www.bbc.com/news/world-europe-34131911

The massive influx of refugees has destabilized Europe politically and culturally, creating divisions between countries and within societies. The displacement of Syria, which is rooted in drought, is erased from the public mind by blaming religion for the upheaval. This is how water wars mutate into culture wars and religious conflicts.

In Nigeria, Boko Haram is presented as an extremist religious movement. However, as Luc Gnacadja, the former head of the UNCCD has attested, "the depletion of Lake Chad helped create the conditions for conflict. In much of northern Nigeria, Muslim herders are in competition with Christian farmers for dwindling water supplies. The so-called religious fight is actually about access to vital resources.

It is not just about Boko Haram, in the Sahel belt, you will see almost the same challenge in Mali and in Sudan. Furthermore, men who were or would have been gainfully employed as farmers, fishermen, fish sellers, and pastoralists have now been conscripted into Boko Haram, with many of them participating in the deadly night raids of the terrorist group. Without a minimum of security of access to the land, restoration of land through investment is not possible. Peace is a prerequisite."

The disappearance of Lake Chad and the emergence of conflicts are interconnected. The Lake Chad basin is one of the most important agricultural heritage sites in the world, providing a lifeline to nearly 30 million people in four countries: Nigeria, Cameroon, Chad, and Niger. Lake Chad gets the name *Chad* from a local word meaning "large expanse of water." It gave its name to the country of Chad. It is the remnant of a former inland sea, paleolake Mega-Chad, which at its largest covered an area of 1,000,000 square kilometers sometime before 5000 BC.

Since the 1960s, Lake Chad has shrunk considerably. In 1983, Lake Chad was reported to have covered 10,000 square kilometers. By the year 2000, it had shrunken to less than 1,500 square kilometers.

The UN Food and Agriculture Organization (FAO) has called the situation an "ecological catastrophe."

The UN Environment Programme (UNEP) and the Lake Chad Basin Commission (LCBC), a regional body that regulates the use of the basin's water and other natural resources, maintain that the damming of rivers and irrigation methods used by the countries bordering the lake are partly responsible for its shrinkage.

The lake is mainly fed by the Chari River through the Lagone tributary, which used to provide 90 percent of the lake's water. The diversion of water from the Chari River to irrigation projects and dams along the Jama'are and Hadejia Rivers in northeastern Nigeria preventing the lake's recharge. As parts of the lake dry up, most farmers and cattle herders have moved toward greener areas, where they compete for land resources with host communities. Others have migrated to Kano, Abuja, Lagos, and other big cities.

The impact of the drying lake is causing tensions among communities around Lake Chad. There are repeated conflicts among nationals of different countries over control of the remaining water. And conflicts have grown between farmers and pastoralists.

> http://www.un.org/africarenewal/magazine/april-2012
> /africa%E2%80%99s-vanishing-lake-chad#sthash.Z2uvVRJs.dpuf

As people are displaced and insecurities grow, identity is transformed and destroyed. Among these vulnerable cultures and identities, terrorism, extremism, and xenophobia take virulent forms. Vicious cycles of violence and exclusion—cultural, political, and economic—predominate.

These are the roots of the rise of the new terrorism and extremism. This is the new face of Water Wars.

These vicious cycles of violence begin with violence against the Earth, violence against water, and violence against the rights of communities, which are all sustained by water.

To make peace, we need to make peace with water and with the Earth. We need to cultivate our deeper identities as earthlings and as water beings. We need to remember that we are water, soil, seed, and earth.

That is why, at the Paris climate meetings, I joined movements from across the world to plant a Garden of Hope and made a pact to protect the Earth and each other.

http://seedfreedom.info/campaign/pact-for-the-earth/

February, 2016
Delhi, India

Converting Abundance into Scarcity

Water is the matrix of culture, the basis of life. In Arabic, Urdu, and Hindustani it is called *ab. Abad raho* is a greeting for prosperity and abundance. The name India itself is derived from the great river Indus, and India was called the land beyond the Indus.[1] Water has been central to the material and cultural well-being of societies all over the world. Unfortunately, this precious resource is under threat. Although two-thirds of our planet is water, we face an acute water shortage.

The water crisis is the most pervasive, most severe, and most invisible dimension of the ecological devastation of the earth. In 1998, 28 countries experienced water stress or scarcity.[2] This number is expected to rise to 56 by 2025. Between 1990 and 2025 the number of people living in countries without adequate water is projected to rise from 131 million to 817 million.[3] India is supposed to fall into the water stress category long before 2025.[4]

A country is said to be facing a serious water crisis when available water is lower than 1,000 cubic meters per person per year. Below this point, the health and economic development of a nation are considerably hampered. When the annual water availability per person drops below 500 cubic meters, people's survival is grievously compromised. In 1951, the average water availability in India was 3,450 cubic meters per person per year.

By the late 1990s, it had fallen to 1,250 cubic meters. By 2050, it is projected to fall to 760 cubic meters. Since 1970, the global per capita water supply has declined by 33 percent.[5] The decline does not result from population growth alone; it is exacerbated by excessive water use as well. During the last century, the rate of water withdrawal has exceeded that of population growth by a factor of two and one-half.[6]

I have witnessed the conversion of my land from a water-abundant country to a water-stressed country. I saw the last perennial stream in my valley run dry in 1982 because of the mining of aquifers in catchments. I have seen tanks and streams dry up on the Deccan plateau as eucalyptus monocultures spread. I have witnessed state after state pushed into water famine as Green Revolution technologies guzzled water. I have struggled with communities in water-rich regions as pollution poisoned their water sources. In case after case, the story of water scarcity has been a story of greed, of careless technologies, and of taking more than nature can replenish and clean up.

The Ecology of Water

The hydrological cycle is the ecological process through which water is received by the ecosystem as rain or snow. The falling moisture recharges streams, aquifers, and groundwater sources. The water endowment of a particular ecosystem depends on the region's climate, physiography, vegetation, and geology. At each of these levels, modern humans have abused the earth and destroyed its capacity to receive, absorb, and store water. Deforestation and mining have destroyed the ability of water catchments to retain water. Monoculture agriculture and forestry have sucked ecosystems dry. The growing use of fossil fuels has led to atmospheric pollution and climate change, responsible for recurrent floods, cyclones, and droughts.

Industrial Forestry and the Water Crisis

Forests are natural dams, conserving water in catchments and releasing it slowly in the form of streams and springs. Rainfall or snowfall is intercepted by forest canopies that protect the soil and increase the potential of forest floors to absorb water. Some of this water evaporates back to the atmosphere. If forest floors are covered with leaf litter and humus, they retain and regenerate water. Forest logging and monoculture agriculture allow water to run off and destroy the water conservation capacity of soils.

Cherapunji in northeast India is the wettest region on earth, with 11 meters of rainfall a year. Today, its forests are gone, and Cherapunji has a drinking-water problem. My own transition from physics to ecology was spurred by the disappearance of Himalayan streams in which I played as a child. The Chipko movement was also launched to stop the destruction of water resources through logging in the area.[7]

The ecological crisis in the Himalayas was accelerated by commercial forestry. Villages once self-sufficient in food production were forced to import food when water sources dried up. With the forests gone, floods and landslides became frequent; in 1970, the Alaknanda disaster, where a major landslide blocked the Alaknanda River and inundated 1,000 kilometers of land, washed away numerous bridges and roads. In 1978, the Tawaghat tragedy took an even greater toll; an entire mountain slope collapsed into the Bhagirathi River, forming a lake four kilometers wide. The lake burst and flooded the Gangetic Plain.[8] The incident was a wake-up call to the government regarding the value of forest catchments.

Long before these flood disasters, there had been warnings about the Himalayan threat. In 1952, Gandhi's disciple Mira Behn remarked:

> Year after year the floods in the north of India seem to be getting worse, and this year they have been absolutely devastating. This means that there is something radically wrong in the Himalayas, and that "something" is, without doubt,

connected with the forests. It is not, I believe, just a matter of deforestation as some people think, but largely a matter of a change
of species.

Living in the Himalayas as I have been continuously now for
several years, I have become painfully aware of a vital change
in species of trees which is creeping up and up the southern
slopes—those very slopes which let down the flood waters on to
the plains below. This deadly changeover is from Banj (Himalayan oak) to Chir pine. It is going on at an alarming speed, and because it is not a matter of deforestation, but of change from one
kind of forest to another, it is not taken sufficiently seriously. In
fact the quasi-commercial Forest Department is inclined to shut
its eyes to the phenomenon, because the Banj brings in no cash
for the coffers, whereas the Chir pine is very profitable.[9]

Despite the value of the leaf litter of oak forests as the primary
mechanism for water conservation in the Himalayan mountain
watersheds, and despite warnings about the disappearance of
the forests, industrial forestry continued unabated, leading to
massive catastrophe in the region.

Eucalyptus and Water Scarcity

In India and other parts of the Third World, the spread of eucalyptus monocultures for the paper and pulp industry has been a
major source of water problems. Eucalyptus, ecologically adapt-
ed to its native habitat in Australia, is hazardous in water-defi-
cient regions. Nowhere outside its native habitat is eucalyptus a
self-sustaining system of vegetation. A study conducted by the
hydrological division of the Australian Central Scientific and
Industrial Research Organization found that during years with
precipitation less than 1,000 millimeters, deficits in soil moisture
and groundwater were created by eucalyptus.[10] Even throughout Australia, reports confirm the rapid destruction of water resources as a consequence of large-scale planting of eucalyptus.

Mahashweta Devi described the impact of eucalyptus on water resources in the tribal areas of Bihar and west Bengal in India:

> I am concerned with the India I know. My India is of the poor, starving, and helpless people. Most of them are landless and the few who have land are happy to be able to make the most of the given resources. To cover Purulia, Bankura, Midnapur, Singbhum, and Palamau with eucalyptus will be to rob my India of drinking and irrigation water.[11]

In 1983, farmers in the state of Karnataka marched en masse to the forest nursery and uprooted millions of eucalyptus seedlings and planted tamarind and mango seeds in their place.[12] In South Africa, women launched a major water campaign to cut down eucalyptus trees that had dried up streams and groundwater sources. South Africa's Working for Water program, spearheaded by the Department of Water Affairs and Forestry, was established to rejuvenate water resources by getting rid of alien plants like eucalyptus, which have invaded more than 10 million hectares and use 3.3 billion cubic meters of water in excess of native vegetation. Shortly after the clearing of eucalyptus along river banks, stream flow increased by 120 percent.[13]

Mining and the Water Crisis

Mining is a practice that destroys water catchments. In the 1980s, limestone mining destroyed my home, Doon Valley. The mining companies saw limestone purely as a raw material for industry; the value of the deep cavities, nature's water reservoirs, was completely ignored. Building an artificial structure with the depth of the Doon Valley catchments would have cost $500 million.[14] In addition to devastating water resources, mining on the precipitous slopes was also causing landslides and filling streams and rivers with debris. I have seen deep and narrow streams transformed into rivers of debris, with beds higher than the surrounding land. Limestone quarrying converted a valley with abundant rainfall into a water-deprived region.

During the conflict over limestone quarrying in Doon Valley, the water resources recharged by the Mussoorie Hills were
treated as worthless and given no consideration. The devaluation of Doon Valley's natural resources was merely an extension
of the devaluation of nature by conventional economics and development models. The failure of modern economics to address
natural resources in their ecological totality has been noted by
many. Nicholas Georgescu-Roegen eloquently summarized this
incompetence of conventional economics:

> The no-deposit, no-return analogy benefits the businessman's
> view of economic life, For, if one looks only at money, all one can
> see is that money just passes from one hand to another: except
> by regrettable accident it never gets out of the economic process.
> Perhaps the absence of any difficulty in securing raw materials
> by those countries where modern economics grew and flour-
> ished was yet another reason for economists to remain blind to
> this crucial economic factor. Not even the wars the same nations
> fought for the control of the world's natural resources awoke the
> economists from their slumber.[15]

The deepening ecological crisis, however, is making it imperative that nature's values and functions be taken into account
through proper ecological audits that assign value to natural
functions on the basis of the cost of technological alternatives to
deliver the same set of goods and services. Thus the value of the
Mussoorie Hills and their potential for water provision would
be equivalent to the cost of technical installations required to
provide the same quantity and quality of water. Quite obviously,
the damage involved is equivalent to the destruction of a gigantic waterworks. Recognizing the social and ecological value of a
resource leads to its equitable and sustainable use. In contrast,
assessing a resource only in terms of market price creates patterns of nonsustainable and inequitable use.

In 1982, the Indian Ministry of Environment in New Delhi invited me and a team of ecologists to conduct an impact assess-

ment of mining. We worked with local communities to build a movement to save the mountains and streams, and we supported citizen groups. The environment ministry initiated legal action to stop limestone mining in Doon Valley, and in 1985 the Supreme Court ordered the permanent or temporary closure of 53 out of the 60 limestone quarries in the region. The court opined:

> This is the first case of its kind in the country involving issues related to environment and ecological balance, and the questions arising for considerations are of grave moment and of significance not only to the people residing in the Mussoorie Hill range forming part of the Himalayas but also in the implications to the welfare of the generality of the people living in the country. It brings into sharp focus the conflict between development and conservation and serves to emphasize the need for reconciling the two.[16]

The court further held that the closure of mining operations was

> a price that has to be paid for protecting and safeguarding the right of the people to live in a healthy environment with minimum disturbance of ecological balance and without avoidable hazards to them and to their cattle, homes and agricultural land and undue affection of air, water and environment.[17]

The decision by the Supreme Court of India was the precedent for accepting a stable and healthy environment as a human right. The court intervened on behalf of citizens.

Unfortunately, globalization is reversing the democratic and ecological victories of the 1980s. Mining is spreading in the most vulnerable areas, including Rajasthan, home to several ancient water systems. Limestone mining has intensified in the coastal regions of Gujarat. Around Gandhi's birthplace, 25 cement factories are scooping out nature's storage and protection systems and exposing the region to water famines. The forests in the sacred Gandmardhan Mountains are a refuge for various plants and provide water to 22 streams, which in turn fill major rivers.

In 1985 the Bharat Aluminum Company (BALCO) began the desecration of these sacred grounds. BALCO was involved in bauxite mining. The company arrived in Gandmardhan after destroying the sanctity and ecology of Amarkantak, another important mountain where the Narmada, Sone, and Mahanadi Rivers originate. Since 1985, the tribals of the region have obstructed the work of the company and refused to be tempted by its offers of employment. Even the police have failed to stop the determined protesters. "Mati Devata, Dharam Devata" ("The soil is our goddess; it is our religion") were words chanted by the women of the "Save Gandmardhan" movement as they were being dragged away by the police. Dhanmati, a 70-year-old protester, summarizes the conviction of the women: "We will sacrifice our lives, but not Gandmardhan. We want to save this hill which gives us all we need."[18]

BALCO's quest for aluminum in this sacred land is particularly disturbing when one considers India's accumulated surplus of the mineral. Local residents have long known how to make aluminum using methods that predate industrial society. Even today, such craftspeople can be found in Orissa. Tribal technology did not destroy the rivers and mountains as industrial mining does. BALCO's mining activity is not based on the needs of the Indian people—it is entirely driven by the demands of industrialized countries whose own aluminum plants are closing for environmental reasons. Japan has reduced its aluminum smelting capacity from 1.2 million tons to 140,000 tons and now imports 90 percent of its aluminum.[19] The survival of Gandmardhan's tribals is under threat because rich countries want to preserve their economies, environments, and luxurious lifestyles.

The local and national ecology movements had stopped mines in many vulnerable catchments to protect rivers. Globalization, however, is reversing many laws. Thirteen minerals—iron, manganese, chrome, sulfur, gold, diamond, copper, lead, zinc, molybdenum, tungsten, nickel, and platinum—have been cleared for exploitation, and mining operations have been deregulated. Automatic approval is given for foreign companies that own 50 percent of the mines. The normal area limit of 25 square kilometers has now been relaxed to 5,000 square kilometers for a single prospecting license.[20]

Large corporations such as Rio Tinto-Zinc (RTZ) are now in Gandmardhan where the local tribals do not want them. As Basano Dehury, an elected representative of her village, points out, "If the company comes, they will dump all the waste and it will block the source of our rivers. Therefore, we do not want the mine"[21] Tikayat Dehury, another villager, wonders, "Why should we work in the mines? We already have what we want. If we work there, it will be we that have to work and work and work and they will take out the cream from here and go."[22]

In Orissa, mining has unleashed a life-and-death battle between local communities and global corporations supported by the military. In December 2000, protesters were killed during an anti-mining demonstration.[23] Whether it is industrial fisheries or forestry, mining or pollution, corporations have stopped the destruction of water resources only when forced by citizens through direct action or through courts.

Drought: An Unnatural Disaster

Since the 1950s, the Green Revolution has been hailed for its success in expanding the global food supply, particularly in developing nations such as India and China.[24] High-yield miracle seeds were promoted all over the developing world, and the Green Revolution was praised for preventing the starvation of millions of people. The ecological and social costs of the Green Revolution were largely ignored. Through its emphasis on high-yield seeds, this agricultural model displaced drought-resistant local crop varieties and replaced them with water-guzzling crops. The water-intensive Green Revolution led to water mining in waterscarce areas.

Prior to the Green Revolution, groundwater was accessed through protective, indigenous irrigation technologies. However, these technologies, which relied on renewable human or animal energy, were identified as "inefficient" and were subsequently replaced by oil engines and electric pumps that extracted water faster than nature's cycles could replenish the groundwater.

Tube Wells and Energized Pumps

Across India, fossil-fuel and electricity-run wells have mush-roomed as part of an informal privatization of groundwater. After the 1972 drought in Maharashtra, the World Bank heavily subsidized and mechanized water withdrawal systems. The bank also gave credit for tube wells that were to feed commercial irri-gation and reduce water scarcity. The result was an explosion of sugarcane cultivation. Maharashtra is now known as the "Land of Sugar Barons." It has recently been discovered that this power was built on the water resources of rural Maharashtra.

In less than a decade, sugarcane fields converted groundwa-ter into a commodity and left people and staple food crops thirst-ing for water. While sugarcane is cultivated on only three percent of Maharashtra's irrigated land, it consumes 80 percent of all the irrigation water and eight times more water than other irrigated crops.[25] As the state struggles with famine, the sugarcane plan-tations and sugar mills flourish. Ten years ago, Maharashtra was home to 77 sugar cooperatives, whose water came from 70 per-cent of the villages. The sugar factories have been actively sup-porting tube well construction. In the meantime, public wells and shallow wells owned by small farmers have run dry.

In the Sangli district, for instance, groundwater irrigation of sugarcane has increased dramatically over the past two decades, even as water scarcity has grown. Although the shift from rain-fed, coarse-grain production to a water-hungry cash crop has increased average household income, the costs have been great. Manerajree village is a perfect example of an area that benefitted

financially in the short run but paid dearly materially and ecologically in the long run. A new water scheme with a potential
supply of 50,000 liters was commissioned in November 1981 at a
cost of $14,000. The water supply lasted only one year. To increase
production, three 60-meter power pump bores were drilled near
the first well, and they supplied 50,000 liters per day in 1982. By
November 1983, all three bores were completely dry. More than
2,000 privately owned wells in this sugarcane region had also
gone dry. Since 1983, there has been a continuous tanker service
providing water to the area.

The Malwa plateau of central India is another tragedy. What
was once a water-abundant region—"Malwa's soil is so rich that
there is food in every home, and water at every step" was a common expression—is now dry, and residents travel an average
distance of four kilometers in search of water. The crisis is a result of dependence on tube wells and the desertion of traditional
water-harvesting systems.

In the village of Belawati, 500 tube wells were created over
the past decade and only five still work.[26] The rest have run dry.
In Guraiya village, only 10 of the 100 tube wells built have water.
In Ismailkhada village, the 1,000 tube wells drilled over a span
of seven years dried up the 12 ponds that served the community
for centuries. Residents now travel two kilometers for water. Of
the 200 tube wells dug in Sadipura, only four are working.[27]

Mechanized water extraction has also created ecological
stress in other parts of the world. Development projects in arid
sub-Saharan Africa played critical roles in the Sahelian famine
of the 1970s and 1980s.[28] Well digging was believed to be the best
mechanism for developing pastoral regions. The traditional pastoral practice of moving herds to different locations was eroded with the introduction of energized wells. The new wells supplied more water than the pastoralists needed and encouraged
their settlement in one location, increasing grazing pressure on
the land. Settling pastoralists in fact worsened the problem of

desertification; it bypassed century-old traditions that ensured survival under conditions of low water availability.

Community Rights and Collective Management

In most indigenous communities, collective water rights and management were the key for water conservation and harvesting. By creating rules and limits on water use, collective water management ensured sustainability and equity. With the advent of globalization, however, community control of water is being eroded, and private exploitation of water is taking hold. Water-renewing traditional systems are now decaying. In a study of 152 villages using traditional water-harvesting systems, 79 were dry or polluted.[29] The Chobala Pond in Mundlana village is still communally maintained, and it still serves the water needs of 10 villages. On the other hand, Mankund, named after the hundreds of ponds and tanks it once boasted, now has no water. The 1,000 tube wells introduced to the region have drained the traditional water sources.[30]

Water is available only if water sources are regenerated and used within limits of renewability. When development philosophy erodes community control and instead promotes technologies that violate the water cycle, scarcity is inevitable. In India, even as capital investment was being poured into water projects, more and more villages were running out of water.

In 1972, the government identified 150,000 villages as facing water problems and introduced programs to provide water in 94,000 of them. The programs included drilling tube wells and pumps to bring water from long distances. Despite these efforts, the number of water-stressed villages had risen to 231,000 by 1980. The government then decided to rescue 94,000 more villages; in 1985, a total of 161,722 villages still faced water problems. More investments were made that year to assist all but 70 villages; but by 1994, 140,975 villages were without water.[31]

In the 1970s and 1980s, the World Bank and other aid agencies focused on disastrous technologies as a means of water provision. Since the 1990s, these agencies have been aggressively pushing privatization and market-based distribution of water, which already promise to be equally catastrophic. In the Indian states of Gujarat and Maharashtra, the World Bank is pushing privatization as a replacement for its own failed technology-intensive water system from the 1980s. The result has been an accelerated extraction of groundwater. In the water-stressed state of Gujarat, groundwater is mined from a depth of 1,500 to 1,800 feet, leaving aquifers shallower and surface storage empty.

Gujarat was once home to a number of highly functional tanks and wells. In the 1930s, wells provided the water for 78 percent of the irrigation in the region.[32] Water was lifted from a well by *kos*, indigenous water-lifting tools, and energy for the wells was provided by animals. When the state was hit with a water crisis in 1985 and 1986, the government, along with the World Bank, created an emergency program, and Gujarat received potable water by special trains, tankers, camels, and bullock carts.

The close to $18 million government program aggravated the problem further. The new sources, including some 4,000 tube wells, ran dry. The government spent an additional $19.4 million on long-distance transfer and on more tube wells. The World Bank also funded a $28.4 million water supply project. In the end, these programs failed to provide water. In fact, the schemes ended up depleting water sources.[33]

The water famine in Maharashtra in the 1980s also reveals a similar story. Ninety-three percent of Maharashtra is made up of hard rocks comprising the Deccan Trap. The Deccan recharge is slow because there is very little storage space for groundwater. In the Deccan Trap, therefore, there is nothing like a subsoil water table. Water is stored in joints and bedding planes and recharged locally. Traditionally, groundwater extraction in Maharashtra came from open dug wells. Fifty-nine percent of the state had been irrigated by groundwater through 939,000 open dug wells. Large-scale development projects have tried to overcome this limitation by digging deeper and using more power for the with-drawal of water. The old methods of withdrawal were regarded as inefficient. As one expert comments:

> There were 5.42 lakh wells in Maharashtra in 1960–61. This num-ber increased to 8.16 lakh in 1980. The average increase per year during the last two decades was 13,700. It is significant to note that although the number of wells increased by about 51 per cent during the 20 years, the area irrigated by them has nearly dou-bled during the same period of years. This is mainly due to the fact that more and more wells are being fitted with mechanised pumps (oil engines and electric pump sets), discarding the out-moded device of draft-like mbots, Persian wheels, etc. Mechani-sation of draft has increased the utility of wells and has resulted in optimum use of water, available for each well.[34]

The notion of increasing well efficiency through energized pumps was short-lived. Powerful water-withdrawal technolo-gies merely led to the exhaustion of water and not to its opti-mum use. The result was groundwater famine.

Ecological Democracy

Technological solutions to an ecological problem have been un-successful. Reductionist assumptions about water development hold that when it comes to using natural resources, nature is deficient and people's traditions are inefficient. However, differ-ent ecozones have been the basis of diverse cultures and econo-mies. The arid zones have been sustainably used for pastoralism, and the semiarid zones have been used for dry farming with protective irrigation. Everyone agrees that the world is facing a severe water crisis. Water-abundant regions have become water scarce, and waterscarce regions face water famines. There are, however, two conflicting paradigms for explaining the water crisis: the market paradigm and the ecological paradigm. The market paradigm sees water scarcity as a crisis resulting from the absence of water trade.

If water could be moved and distributed freely through free markets, this paradigm holds, it would be transferred to regions of scarcity, and higher prices would lead to conservation. As Anderson and Snyder state, "[A]t higher prices people tend to consume less of a commodity and search for alternative means of achieving their desired ends. Water is no exception."[35]

Market assumptions are blind to the ecological limits set by the water cycle and the economic limits set by poverty. Overexploitation of water and disruption of the water cycle create absolute scarcity that markets cannot substitute with other commodities. The assumption of substitution is in fact central to logic of commodification. For example, economist Jack Hirshleifer and his colleagues state:

> This is not to deny that as a commodity, water has its special features, for example, its supply is provided by nature partly as a store and partly as a flow, and it is available without cost in some locations but rather expensive to transport to others. Whatever reason we cite, however, the alleged unique importance of water disappears upon analysis.[36]

Such abstract arguments miss the most crucial point—when water disappears, there is no alternative. For Third World women, water scarcity means traveling longer distances in search of water. For peasants, it means starvation and destitution as drought wipes out their crops. For children, it means dehydration and death. There is simply no substitute for this precious liquid, necessary for the biological survival of animals and plants.

The water crisis is an ecological crisis with commercial causes but no market solutions. Market solutions destroy the earth and aggravate inequality. The solution to an ecological crisis is ecological, and the solution for injustice is democracy. Ending the water crisis requires rejuvenating ecological democracy.

1. Bill Aitkin, *Seven Sacred Rivers* (Columbia, MO: South Asia Books, 1992), p. 1.
2. Marq De Villiers, *Water: The Fate of Our Most Precious Resource* (New York: Houghton Mifflin, 2000), p. 17.
3. Ibid., p. 18.
4. Robin Clarke, *Water: The International Crisis* (Cambridge, MA: MIT Press, 1993), p. 67.
5. Sandra Postel, *Water for Agriculture* (Washington, DC: Worldwatch Institute, 1989).
6. Ibid.
7. Vandana Shiva, *Staying Alive: Women, Ecology and Development in India* (London: Zed Books, 1988), pp. 67-77.
8. Vandana Shiva et al. *Ecology and the Politics of Survival: Conflicts Over Natural Resources in India* (New Delhi: Sage, 1991), p. 109.
9. Mira Behn, "Something Wrong in the Himalaya," (n.d.)
10. Vandana Shiva et al., *Ecology and the Politics of Survival*, p. 147.
11. Ibid.
12. Vandana Shiva, *Staying Alive*, p, 82.
13. Personal communication, Kader Asmal, Water Minister, South Africa; CSIR Division of Water, Environment and Forestry Technology, *The Environmental Impacts of Invading Alien Plants in South Africa* (Pretoria, SA: Department of Water Affairs and Forestry, 2001).
14. Vandana Shiva et al., *Doon Valley Ecosystem* (Government of India: Report produced for the Ministry of Environment).
15. Nicholas Georgescu-Roegen, *The Entropy Law and the Economic Process* (Cambridge, MA: Harvard University Press, 1974), pp. 2-21.
16. Shiva et al., *Ecology and the Politics of Survival*, p. 300.
17. Ibid.
18. Vandana Shiva, "Homeless in the Global Village," in Maria Mies and Vandana Shiva, *Ecofeminism* (Halifax, NS: Fernwood Publications; London: Zed Books, 1993), p. 100.
19. Vandana Shiva and Afsar Jafri, *Stronger than Steel: People's Movement Against Globalisation and the Gopalpur Steel Plant* (New Delhi: Research Foundation for Science, Technology, and Ecology, 1998), p. 1.
20. Vandana Shiva et al., *The Ecological Costs of Globalisation* (New Delhi: Research Foundation for Science, Technology and Ecology, 1997), p. 7.
21. "What is RTZ Doing in Orissa?" (report by Mines, Minerals and People, April 15,2001).
22. Ibid.
23. Prafulla Samantra, "Kashipur Alumina Projects and the Voice of Tribals for Life and Livelihood," (presentation at the. Conference on Globalisation and Environment sponsored by the Research Foundation for Science, Technology and Ecology, September 30, 2001).

24. Vandana Shiva, *Violence of the Green Revolution* (London: Zed Books, 1991).
25. V. B. Vebalkar, "Irrigation by Groundwater in Maharashtra," (Poona, India: Groundwater Survey and Development Agency, 1984).
26. Anjana Trivedi and Rajendar Bandhu, "Report of Water Scarcity in Malwa," *Niti Marg* (May 2001), pp. 19-25.
27. Ibid.
28. Lloyd Timberlake, *Africa in Crisis: The Causes, the Cures of Environmental Bankruptcy* (London: International Institute for Environment and Development, 1985).
29. Anjana Trivedi and Rajendar Bandhu, "Report of Water Scarcity in Malwa."
30. Ibid.
31. Centre for Science and Environment, "Water Report," Delhi, 2000.
32. Vandana Shiva et al., *Ecology and the Politics of Survival*, p. 187.
33. "Gujarat in for Acute Water Famine," *Times of India*, December 20,1986; "Solutions that Hold No Water," *Times of India*, December 8, 1986.
34. V. B. Hebalkar, "Irrigation by Groundwater in Maharashtra."
35. Terry Anderson and Pamela Snyder, *Water Markets: Priming the Invisible Pump* (Washington DC: Cato Institute, 1997), p. 8.
36. Jack Hirshleifer, James C. De Haven, and Jerome W. Milliman, *Water Supply: Economics, Technology, and Policy* (Chicago, IL: University of Chicago Press, 1960).

CHAPTER ONE

Water Rights:
The State, the Market,
the Community

W ho does water belong to? Is it private property or a commons? What kind of rights do or should people have?
What are the rights of the state? What are the rights of corporations and commercial interests? Throughout history, societies
have been plagued with these fundamental questions.

We are currently facing a global water crisis, which promises
to get worse over the next few decades. And as the crisis deepens,
new efforts to redefine water rights are under way. The globalized economy is shifting the definition of water from common
property to private good, to be extracted and traded freely. The
global economic order calls for the removal of all limits on and
regulation of water use and the establishment of water markets.
Proponents of free water trade view private property rights as
the only alternative to state ownership and free markets as the
only substitute to bureaucratic regulation of water resources.

More than any other resource, water needs to remain a common good and requires community management. In fact, in
most societies, private ownership of water has been prohibited.
Ancient texts such as the *Institute of Justinian* show that water
and other natural sources are public goods: "By the law of nature

these things are common to mankind—the air, running water, the sea, and consequently the shore of the sea."[1] In countries like India, space, air, water, and energy have traditionally been viewed as being outside the realm of property relations. In Islamic traditions, the *Sharia*, which originally connoted the "path to water," provides the ultimate basis for the right to water. Even the United States has had many advocates for water as a common good. "Water is a moving, wandering thing, and must of necessity continue to be common by the law of nature," wrote William Blackstone, "so that I can only have a temporary, transient, usufructuary property therein."[2]

The emergence of modern water extraction technologies has increased the role of the state in water management. As new technologies displace self-management systems, people's democratic management structures deteriorate and their role in conservation shrinks. With globalization and privatization of water resources, new efforts to completely erode people's rights and replace collective ownership with corporate control are under way. That communities of real people with real needs exist beyond the state and the market is often forgotten in the rush for privatization.

Water Rights as Natural Rights

Throughout history and across the world, water rights have been shaped both by the limits of ecosystems and by the needs of people. In fact, the root of the Urdu word *abadi*, or human settlement, is *ab*, or water, reflecting the formation of human settlements and civilization along water sources. The doctrine of riparian right—the natural right of dwellers supported by a water system, especially a river system, to use water—also arose from this concept of *ab*. Water has traditionally been treated as a natural right—a right arising out of human nature, historic conditions, basic needs, or notions of justice. Water rights as natural rights do not originate with the state; they evolve out of a given ecological context of human existence.

As natural rights, water rights are usufructuary rights; water can be used but not owned. People have a right to life and the resources that sustain it, such as water. The necessity of water to life is why, under customary laws, the right to water has been accepted as a natural, social fact:

> The fact that right over water has existed in all ancient laws, including our own *dharmasastras* and the Islamic laws, and also the fact that they still continue to exist as customary laws in the, modern period, clearly eliminates water rights as being purely legal rights, that is, rights granted by the state or law.[3]

Riparian Rights

Riparian rights, based on concepts of usufructuary rights, common property, and reasonable use, have guided human settlement all over the world. In India, riparian systems have long existed along the Himalaya. The famous grand *Anient* (canal) on the Kaveri at the Ullar River dates back a thousand years and is believed to be the oldest hydraulic structure to control the flow of rivers in India. It is still functioning. In the northeast, old riparian systems known as *dongs* guide the use of water. In Maharashtra, conservation structures were known as *bandharas*.

The *ahar* and *pyne* systems of Bihar, where an unlined inundation canal (*pyne*) transfers water from a stream into a catchment basin (*ahar*), also evolved from a riparian doctrine. Unlike modern Sone canals built by the British, which have failed to meet the needs, of the people, the *ahars* and *pynes* still provide water to peasants. In the United States, riparian systems were introduced by the Spanish, who had brought them from the Iberian Peninsula.[4] These systems were adopted in Colorado, New Mexico, and Arizona, as well as the eastern settlements.

Early riparian principles were based on the notion of sharing and conserving a common water source. They were not attached to property rights. As historian Donald Worster notes:

In ancient times, the riparian doctrine was less a method of as-
certaining individual property rights and more the expression
of an attitude of non-interference with nature. Under the oldest form of the principle a river was to be regarded as no one's
private property. Those who lived along its banks were granted
rights to use the flow for natural purposes like drinking, washing,
or watering their stock, but it was a usufructuary right only—a
right to consume so long as the river was not diminished.[5]

Even European colonists who first settled in the eastern
United States adhered to these basic tenets. But as the western
part of the country began to be inhabited, usufructuary rights
were no longer prevalent. The riparian concept was instead believed to have emerged from English common law and consequently centered around individual property ownership. "The
men and women who settled the American West did not belong
to that older world ... [They] rejected the traditional riparran-
ism," writes Worster. "Instead, they chose to set up over most
of the region the doctrine of prior appropriation because it offered them a greater freedom to exploit nature"[6] Universal water
rights were thus severely curtailed.

Cowboy Economics: The Doctrine of Prior Appropriation and the Advent of Privatization

It was in the mining camps of the American West that the cowboy notion of private property and the rule of appropriation—
Qui prior est in tempore, potior est in jure (He who is first in time is
first, in right)—first emerged. The doctrine of prior appropriation
established absolute rights to property, including the right to
sell and trade water. New water markets blossomed and soon replaced natural water rights and the value of water was determined
by the monopolistic first settlers. Prior appropriation "gave no
preference to riparian landowners, allowing all users an opportunity to compete for water and to develop far from streams."[7]

The cowboy sentiment "might is right" meant that the economically powerful could invest in capital-intensive means to appropriate water regardless of the needs of others and the limits of water systems. This frontier logic granted the first appropriator an exclusive right to the water. Latecomers could appropriate water on the condition that prior rights were honored first. Cowboy economics permitted the diversion of water from streams to be used on nonriparian lands. If the appropriator did not use the water, he was forced to forfeit his right.

The cowboy logic allowed the transfer and exchange of water rights among individuals, who often disregarded water's ecological functions or its functions beyond mining. Although rights were based on first settlement, the true first settlers—Native Americans—were denied water appropriation rights. Miners and colonizers, assumed to be the first inhabitants, were granted all rights to use the water sources.[8]

Disregard for the limits of nature's hydrological cycle meant that rivers could be drained and polluted by mining waste. Disregard for the natural rights of others meant that people were denied access to water, and regimes of unequal and nonsustainable water use and water-wasteful agriculture began to spread across the American west.

Contemporary Cowboy Economics

The current push to privatize common water sources had its foundation in cowboy economics. Champions of water privatization, such as Terry Anderson and Pamela Snyder of the conservative Cato Institute, not only acknowledge the link between current privatization efforts and cowboy water laws, but also look at the earlier Western appropriation philosophy as a model for the future:

> From the western frontier, especially the mining camps, came the doctrine of prior appropriation and the foundation of water marketing. This system provided the essential ingredients

for an efficient market in water wherein property rights were well-defined, enforced and transferable.[9]

The current push to reintroduce and globalize the lawlessness of the frontier is a recipe for destroying our scarce water resources and for excluding the poor from their water share. Parading as the anonymous market, the rich and powerful use the state to appropriate water from nature and people through the prior-appropriation doctrine. Private interest groups systematically ignore the option of community control over water. Because water falls on earth in a dispersed manner and because every living being needs water, decentralized management and democratic ownership are the only efficient, sustainable, and equitable systems for the sustenance of all. Beyond the state and the market lies the power of community participation. Beyond bureaucracies and corporate power lies the promise of water democracy.

Water as a Commons

Water is a commons because it is the ecological basis of all life and because its sustainability and equitable allocation depend on cooperation among community members. Although water has been managed as a commons throughout human history and across diverse cultures, and although most communities manage water resources as common property or have access to water as a commonly shared public good even today, privatization of water resources is gaining momentum.

Prior to the arrival of the British in south India, communities managed water systems collectively through a system called *kudimaramath* (self-repair). Before the advent of corporate rule by the East India Company in the 18th century, a peasant paid 300 out of 1,000 units of grain he or she earned to a public fund, and 250 of those units stayed in the village for maintenance of commons and public works.[10] By 1830, peasant payments rose to 650 units, out of which 590 units went straight to the East India Company. As a result of increased payments and lost maintenance revenue, the peasants and commons were destroyed. Some 300,000 water tanks built over centuries in pre-British India were destroyed, affecting agricultural productivity and earnings.

The East India Company was driven out by the first movement for independence in 1857. In 1858, the British passed the Madras Compulsory Labor Act of 1858, popularly known as the Kudimaramath Act, mandating peasants to provide labor for the maintenance of the water and irrigation systems.[11] Because *kudimaramath* was based on self-management and not coercion, the act failed to mobilize community participation and to rebuild the commons.

Self-managed communities have not just been a historical reality; they are a contemporary fact. State interference and privatization have not wiped them out entirely. In a nationwide survey covering districts in dry tropical regions in seven states, N. S. Jodha finds that the most basic fuel and fodder needs of the poor throughout India continue to be satisfied from common property resources.[12] Jodha's studies of commons in the fragile Thar desert also reveal that village community councils still adjudicate grazing rights: institutional rules and regulations determine periods of restricted grazing, the rotational patterns for grazing, the numbers and types of animals to be grazed, the rights to dung and fuel wood collection, and the rules for lopping trees for green fodder. Village councils also appoint their own watchmen to ensure that no community member or outsider breaks the rules. Similar rules exist for maintenance of wells and tanks.

Tragedy of the Commons

John Locke's treatise on property effectively legitimized the theft of the commons in Europe during the enclosure movements of the 17th century. Locke, son of wealthy parents, sought to defend capitalism—and his family's massive wealth—by arguing that property was created only when idle natural resources were transformed from their spiritual form through the application of labor: "Whatsoever, then, he removes out of the state that Nature hath provided and left in it, he hath mixed his labor with it, and joined to it something that is his own, and thereby makes it his property"[13] Individual freedom was dependent upon the freedom to own, through labor, the land, forests, and rivers. Locke's treatises on property continue to inform theories and practices that erode the commons and destroy the earth.

In contemporary times, water privatization is based on Garrett Hardin's *Tragedy of the Commons*, first published in 1968. To explain his theory, Hardin calls on us to imagine a scenario:

> Picture a pasture open to all. It is to be expected that each herdsman will try to keep as many cattle as possible on the commons. Such an arrangement may work reasonably satisfactorily for centuries because tribal wars, poaching, and disease keep the numbers of both man and beast well below the carrying capacity of the land. Finally, however, comes the day of reckoning, that is, the day when the long-desired goal of social stability becomes a reality. At this point, the inherent logic of the commons remorselessly generates tragedy.[14]

Hardin assumes that commons were socially unmanaged, open-access systems with no ownership. And Hardin sees the absence of private property as a recipe for lawlessness.

Although Hardin's theory about the commons has gained tremendous popularity, it is has several holes. His assumption about commons as unmanaged, open-access systems stems from the belief that management takes effect only in the hands of private individuals. But groups do manage, themselves, and commons are regulated rather well by communities. Moreover, commons are not open-access resources as Hardin proposes; they in fact apply the concept of ownership, not on an individual basis, but at the level of the group. And groups do set rules and restrictions regarding use. Regulations of utility are what protect pastures from overgrazing, forests from disappearing, and water resources from vanishing.

Hardin's prediction about the doom of commons has at its center the idea that competition is the driving force in human societies. If individuals do not compete to own property, law and order will be lost. This argument has failed to hold ground when tested in large sections of rural societies in the Third World, where the principle of cooperation, rather than competition, among individuals still dominates. In a social organization based on cooperation among members and need-based production, the logic of gain is entirely different from those in competitive societies. Garrett Hardin's *Tragedy of the Commons* misses the critical point that under circumstances in which common lands cannot even support the basic needs of the population, a tragedy is inevitable—with or without competition.

Communities and Commons

In the upper reaches of the Rio Grande Valley in Colorado, water is still managed as a commons. I had the opportunity to visit San Luis, home of traditional *acequia* systems (gravity-driven irrigation ditch) that nurture soils, plants, and animals. I was there to offer solidarity to the local communities engaged in a major struggle to defend the commons and the oldest system of water rights in Colorado. What the irrigation ditches produce is not merely a market commodity but a denseness of life. "The ditches make a lot of plant life possible in what is really a cold, barren desert," says Joseph Gallegos, a fifth-generation farmer working on ancestral lands in San Luis. "More plants means that the wildlife—birds and mammals—have a home. The ecologists call this biodiversity. I call it life, *terra y vida*."[15]

When the water of the Rio Grande is auctioned to the highest bidder, it is taken away from the agri-pastoral community whose rights to the water are tied to the responsibility of maintaining a "watershed commonwealth."[16] Markets fail to capture diverse values, and they fail to reflect the destruction of ecological value. Water that replenishes ecosystems is considered water wasted. Joseph Gallegos raises an important point when he asks:

> Whose point of view is this? The cottonwood trees that line the acequia banks don't think the leaking water is wasted. Nor do the birds and other animals that live in the trees. The ditches create habitat niches for wildlife, and that is a good thing for the animals and the farmers. It is not wasteful, unless of course you are an urban developer greedily looking for more water for the cities' maniacal growth needs. The gringo treats water like a commodity. You know the saying, "In Colorado water flows uphill, towards money."[17]

When money determines value and courts get involved, common resources are stripped from farmers and lost to private companies. And, as Devon Peña points out,

> The attack on common property rights involves the legal codification of production that produces violent but legally sanctioned invasions, enclosures, and expropriations of space. The law itself violates the integrity of places as habitat for mixed communities of humans and non-humans.[18]

This is exactly what transpired in the Rito Seco Watershed in Colorado, when courts allowed the Batde Mountain Gold Mine to transfer water from agriculture to industrial use.

Community Rights and Water Democracies

Under conditions of scarcity, sustainable systems of water management evolved from the idea of water as commons passed on from generation to generation. Labor in conservation and community building became the primary investment in water resources. In the absence of capital, people working collectively provided the major input or "investment" in water works. As Anupam Mishra of the Gandhi Peace Foundation observes:

> The ways of collecting the drops of Palar, i.e., of rainfall, are as unending as the names of clouds and drops. The pot like the

ocean is filled up drop by drop. These beautiful lessons are not to be found in any textbook but are actually couched in the memory of our society. It is from this memory that the *shrutis* of our oral traditions have come.... The people of Rajasthan did not entrust the organisation of such a boundless work to either the central or federal government, not even to what in modern parlance is termed as the private sphere. It is the people themselves who in each house, in each village, gave fruition to this structure, maintained it, and further developed it.

"Pindwari" is to help others through one's effort, one's labour, one's hard work. The drops of sweat streaming down the brow of the people of Rajasthan continue to flow so as to collect the drops of rain.[19]

Traditional water systems based on local management were insurance against water scarcity in drought-prone regions of Gujarat. These systems were managed mainly by village committees. In the event of floods, famines, and other calamities, the king also helped; the role of a central authority was, therefore, primarily in disaster mitigation. Local institutions in water management included farmers' associations, local irrigation functionaries, local irrigation technicians, the village water associations, and the community labor system, maintained by contributions from each family.

In India, farmers' associations for the construction and maintenance of water systems were once widespread. In Karnataka and Maharashtra the associations were known as panchayats. In Tamil Nadu, they were called *nattamai, kavai maniy am, nir maniyam, oppidi sangam,* or *eri variyam* (tank committee). Tanks and ponds often served more than one village, and in such cases representatives from each village or farmers' association ensured democratic control. These committees could also collect tank dues and taxes from users. Lands were also donated, especially for financing capital expenditures on waterworks.

Village water systems required irrigation functionaries who looked after the day-to-day operation of irrigation systems. In the

Himalayas, where *kuhls* served community irrigation needs, irrigation managers were called *kohlis*. In Maharashtra, they were known as *patkaris*, *havaldars*, and *jogalaya*. In Karnataka and Tamil Nadu, they were known as *nirkatti*, *nirganti*, *nirpaychi*, *niranikkans*, or *kamkukatti*.

To ensure neutrality, *nirkattis* were chosen from the landless caste—the Harijans—who were granted autonomy from landowners and caste groups. Only Harijans held the power to close and open the tanks or vents. Once the farmers laid down the rules of distribution, no individual farmer could interfere, and those who did could be fined. This protection of the associations from the economically powerful ensured water democracy.

Compensations were based on investments of one's own labor and could not be substituted by capital or by others' labor. In South India, collective labor investment was the primary investment in the construction and maintenance of village water systems known as *kudimaramath*. Each able-bodied person was required to help maintain and clean channels. *Nirkattis* also summoned farmers to clean the supply and field channels. The ancient economic treatise, *Arthasastra*, included certain punishments for defaulters from any kind of cooperative construction. Violators were expected to send their servants and bullocks to carry on their work and to share the costs, without laying any claim to the return.

The self-management systems suffered when the government took control over water resources during British rule. Community ownership was further eroded with the emergence of bore wells and tube wells, which made individual farmers dependent on capital. Collective water rights were undermined by state intervention, and resource control was transferred to external agencies. Revenues were no longer reinvested in local infrastructure but diverted to government departments.

Community rights are necessary for both ecology and democracy. Bureaucratic control by distant and external agencies and market control by commercial interests and corporations create disincentives for conservation. Local communities do not conserve water or maintain water systems if external agencies—bureaucratic or commercial—are the only beneficiaries of their efforts and resources.

Higher prices under free-market conditions will not lead to conservation. Given the tremendous economic inequalities, there is a great possibility that the economically powerful will waste water while the poor will pay the price. Community rights are a democratic imperative—they hold states and commercial interests accountable and defend people's water rights in the form of decentralized democracy.

The Right to Clean Water versus the Right to Pollute

Prior to passage of the Water Act of India in 1974, almost all judicial decisions were in favor of polluters. In addition to being protected by law, polluters also had more economic and political power than ordinary citizens. They were even more successful in using the legal processes in their favor. When the impact of industrial pollution was not severe or when industrialization was seen as a symbol of progress, courts tended to uphold the rights of the industrialists to pollute water as exemplified in a number of cases: *Deshi Sugar Mills v. Tups Kahar, Empress v. Holodhan Poorroo; Emperor v. Nana Ram; Imperatix v. Neelappa; Darvappa Queen v. Vittichakkon; Reg v. Partha;* and *Imperatix v. Hari Baput.* As water pollution intensified with the spread of industrialization, it could be handled only through criminal or penal sanctions. However, the courts alone could not protect people's right to clean water.

By the 1980s, as the threat from pollution increased, the right to clean water had to be defended as a fundamental right. The Supreme Court of India introduced a new principle of environmental rights in the famous case *Ratlam Municipality v. Vardhichand.* The municipality had to remove public nuisances, whether it had the financial capability to do so or not. *Ratlam* established a new type of natural right and recognized customary rights as a constitutional guarantee. But even after *Ratlam* and the Water Act, the big polluters were not brought under the law. In most cases, the Central Water Pollution Board was against small factories.[20]

In the industrial world, antipollution regulations were introduced primarily to clean up rivers. In 1969, the Cuyahoga River in Cleveland, Ohio, which served as a dump site for industries, was so contaminated by chemicals that it caught fire. In 1972, the United States passed the Clean Water Act, which established that no one had a right to pollute water and that everyone had a right to clean water. Before the passage of the law, water pollution was handled as a matter of common law involving trespassing and nuisance. The act set the goal of rendering the waters fishable and swimmable by 1983, and eliminating discharges of water pollutants by 1985. Since the passage of the Clean Water Act in 1972, US pollution from point sources has been dramatically reduced, showing the power of regulation in pollution control.

In 1977, as a result of pressure from industry, the focus in the United States shifted from control-point discharge regulation to water quality standards. Tacitly, this shift marked a move away from pollution as a violation to pollution as permissible. Companies attempted to reintroduce the right to pollute through back-door efforts such as tradable pollution rights or tradable discharge permits (TDPs). Although TDPs have faced resistance from environmentalists, they still remain a popular market myth for solving pollution problems.

Supporters of the free market promote TDPs as an alternative to the "command-and-control" of environmental regulation. However, trade in pollution is also government sanctioned. As free-market advocates Snyder and Anderson admit, "Tradable pollution rights are essentially an assignment by a governmental agency of a right to discharge a specified level of pollution into a water body or water course."[21] The government also sets pollution standards, albeit on the basis of a fictitious "bubble," an imagined boundary covering a designated area.

It is not surprising that pollution permits are ecologically blind. They merely consider "incentives for gains from trade." If pollution control costs are low, an industry will sell discharge rights, and if costs are high, an industry will buy discharge rights. While such cost-benefit analysis might appear to create trade advantages, this market of pollution is ecologically dangerous.

Trade in pollution permits violates ecological democracy and people's right to clean water on several counts. It changes the role of governments from protector of people's water rights to advocate of polluters' rights. Governments assume regulatory roles that are anti-environment, anti-people, and pro-polluter industry. TDPs exclude nonpolluters and ordinary citizens from an active democratic role in pollution control, since the trade in pollution is restricted to polluter industries.

Big Polluters: Old and New

The struggle between the right to clean water and the right to pollute is the struggle between the human and environmental rights of ordinary citizens and the financial interests of businesses. Pollution is a byproduct of industrial technologies and global trade. Handmade paper and vegetable dyes cause no pollution; indigenous leather treatment is also very prudent and water conserving; fresh vegetables and fruits do not require water, except for cultivation.

By contrast, modern industrial papermaking and leather processing create massive pollution. Pulp uses 60,000 to 190,000 gallons of water per ton of paper or rayon. Bleaching uses 48,000 to 72,000 gallons of water per ton of cotton. Packaging green beans and peaches for long-distance trade can use up to 17,000 and 4,800 gallons per ton, respectively.[22]

The overuse and pollution of scarce water resources is not restricted to old industrial technologies; it is a hidden component of the new computer technologies. A study by South West Network for Environmental and Economic Justice and the

Campaign for Responsible Technology reveals that the process of chip manufacturing requires excessive amounts of water.

On average, processing a single six-inch silicon wafer uses 2,275 gallons of deionized water, 3,200 cubic feet of bulk gases, 22 cubic feet of hazardous gases, 20 pounds of chemicals, and 285 kilowatts hours of electrical power.[23] In other words,

> if an average plant processes 2,000 wafers per week (the new state-of-the-art Intel facility in Rio Rancho, New Mexico, for example, can produce 5,000 wafers per week) it would need 4,550,000 gallons of water per week and 236,600,000 gallons per year for wafer production alone.[24]

The study finds that out of the 29 Superfund sites in Santa Clara County, California, 20 were created by the computer industry.

The Principles of Water Democracy

At the core of the market solution to pollution is the assumption that water exists in unlimited supply. The idea that markets can mitigate pollution by facilitating increased allocation fails to recognize that water diversion to one area comes at the cost of water scarcity elsewhere.

In contrast to the corporate theorists who promote market solutions to pollution, grassroots organizations call for political and ecological solutions. Communities fighting high-tech industrial pollution have proposed the Community Environmental Bill of Rights, which includes rights to clean industry, to safety from harmful exposure, to prevention, to knowledge, to participation, to protection and enforcement, to compensation, and to cleanup.[25] All of these rights are basic elements of a water democracy in which the right to clean water is protected for all citizens. Markets can guarantee none of these rights.

There are nine principles underpinning water democracy:

1. Water is nature's gift

We receive water freely from nature. We owe it to nature to use this gift in accordance with our sustenance needs, to keep it clean and in adequate quantity. Diversions that create arid or waterlogged regions violate the principles of ecological democracy.

2. Water is essential to life

Water is the source of life for all species. All species and ecosystems have a right to their share of water on the planet.

3. Life is interconnected through water

Water connects all beings and all parts of the planet through the water cycle. We all have a duty to ensure that our actions do not cause harm to other species and other people.

4. Water must be free for sustenance needs

Since nature gives water to us free of cost, buying and selling it for profit violates our inherent right to nature's gift and denies the poor of their human rights.

5. Water is limited and can be exhausted

Water is limited and exhaustible if used nonsustainably. Nonsustainable use includes extracting more water from ecosystems than nature can recharge (ecological nonsustainability) and consuming more than one's legitimate share, given the rights of others to a fair share (social nonsustainability).

6. Water must be conserved

Everyone has a duty to conserve water and use water sustainably, within ecological and just limits.

7. Water is a commons

Water is not a human invention. It cannot be bound and has no boundaries. It is by nature a commons. It cannot be owned as private property and sold as a commodity.

8. No one holds a right to destroy

No one has a right to overuse, abuse, waste, or pollute water systems. Tradable pollution permits violate the principle of sustainable and just use.

9. Water cannot be substituted

Water is intrinsically different from other resources and products. It cannot be treated as a commodity.

1. *Institutes of Justinian* 2.1.1
2. William Blacks tone, quoted in Walter Prescott Webb, *The Great Plains* (New York: Grosset and Dunlop, 1931).
3. Chattarpati Singh, "Water and Law" (n.d.).
4. Devon Pena, ed., *Chicano Culture, Ecology and Politics* (Tucson, AZ: University of Arizona Press, 1998), p. 235.
5. Donald Worster, *Rivers of Empire: Water; Aridity, and the Growth of the American West* (New York: Pantheon Books, 1985), p. 88.
6. Ibid., p.89
7. Ibid., p.104.
8. Ibid., p.90.
9. Terry Anderson and Pamela Snyder, *Water Markets: Priming the Invisible Pump* (Washington, DC: Cato Institute, 1997), p. 75.
10. Jatinder Bajaj, "Green Revolution: A Historical Perspective" (paper presented at 'CAP/TWN Seminar on "Crisis of Modern Science," Penang, November 1986), p. 4.
11. Nirmal Sengupta, *Managing Common Property: Irrigation in India and The Philippines* (New Delhi: Sage, 1991), p. 30.
12. N. S. Jodha, "Common Property Resources'and Rural Poor," *Economic and Political Weekly* 21, No. 7 (July 5, 1986).
13. John Lock, *Second Treatise on Civil Government*, (Buffalo, NY: Prometheus Books, 1986), p. 20.
14. Garrett Hardin, "Tragedy of the Commons," Science 162 (1968): pp. 1243-1248.
15. Devon Peña, ed., *Chicano Culture, Ecology and Politics* (Tucson, AZ: University of Arizona Press, 1998), p. 235.
16. Ibid.
17. Ibid., p. 242.
18. Devon Peña, "A Gold Mine, an Orchard, and an Eleventh Commandment," in Pena ed., *Chicano Culture, Ecology and Politics* (Tucson, AZ: University of Arizona Press, 1998), pp. 250-251.
19. Anupam Mishra, "The Radiant Raindrops of Rajasthan," translated by Maya Jani (New Delhi: Research Foundation for Science, Technology and Ecology, 2001).
20. Chattarpati Singh, "Water and Law."
21. Terry Anderson and Pamela Snyder, *Water Markets*, p. 149.
22. Peter Rogers, *Americas Water: Federal Roles and Responsibilities* (Cambridge, MA: MIT Press, 1993).
23. South West Network for Environmental and Economic Justice and Campaign for Responsible Technology, *Sacred Waters* (1997), pp. 19-20.
24. Ibid.
25. Ibid., pp. 133-134.

Ryots irrigating rice fields.

CHAPTER TWO

Climate Change
and the Water Crisis

"Jala bahule srustinasa, jala bihune srustinasa."
("Too much or too little water destroys creation.")

—Oriya expression

In October 1999, a killer cyclone hit the eastern part of the state of Orissa in eastern India. The cyclone, one of the most devastating human disasters ever experienced, damaged 1.83 million houses and 1.8 million acres of paddy crops in 12 coastal districts. Eighty percent of the coconut trees were uprooted or broken in half, and all the banana and papaya plantations were wiped out. More than 300,000 cattle perished, more than 1,500 fishermen and fisherwomen lost their entire source of livelihood, and more than 15,000 ponds were contaminated or salinated. While there is no official number of the human casualties, independent observers and local workers estimate the toll to be about 20,000.

In the summer of 2001, Orissa was hit by one of the worst droughts in history, and during the monsoon season it was affected by the worst flood. More than seven million people were affected: 600,000 villages were marooned, 42 people were killed, and 550,000 hectares of standing crops were destroyed. Heavy rains in the Mahanadi catchment had forced the release of 13 million cubic meters per second of waters from the Hirakud Dam.

Water is life, but too much or too little of it can become a threat to life. The stories of Noah and Vishnu Purana are tales of mythic floods that wiped out life on the planet. While floods and droughts have always occurred, they have become more intense and more frequent. These climatic extremes are linked to climate change, which is in turn linked to pollution of the atmosphere by the use of fossil fuels.

Climate Injustice as Water Injustice

The impact of climate crisis on all forms of life is mediated through water in the form of floods, cyclones, heat waves, and droughts. Water fury can be tamed only if the atmospheric saturation by carbon dioxide is contained. While subverting international struggle to avert climate disaster makes economic sense for oil companies, it spells political and ecological disaster for much of the earth's community. More than anything, the oil economy's environmental externalities, such as atmospheric pollution and climate change, will determine the future of water, and through water, the future of all life.

Climate destabilization, although set in motion with the advent of industrialization, did not accelerate until very recently. In 1850, the global carbon dioxide in atmosphere was roughly 280 parts per million (ppm); by the mid-1990s, it had increased to approximately 360 ppm.[1] Climate instability—in the form of more extreme floods and droughts, more frequent heat waves and freezing winters—is the result of atmospheric pollution aggravated by the wealthier regions of the world, Since 1950, 11 countries have contributed 530.3 billion tons of carbon dioxide. Of that, the United States has contributed 186.1 billion; the European Union, 127.8 billion; Russia, 68.4 billion; China, 57.6 billion; Ukraine, 21.7 billion; India, 15.5 billion; Canada 14.9 billion; Poland, 14.4 billion; South Africa, 8.5 billion; Mexico, 7.8 billion; and Australia, 7.6 billion.

As the level of carbon dioxide increases, these molecules trap more heat and global temperatures rise. Along with other greenhouse gases, such as methane and nitrogen, the impact of carbon dioxide promises to be catastrophic. Methane concentration, for instance, has risen from 0.7 parts per million four centuries ago to 1.7 parts per million in 1988.[2] About 10 percent of the feed given to animals in industrial livestock factories goes into the atmosphere as methane.[3] This gas is also responsible for the foul smell surrounding factory farms.

In May 1988, 50 countries held the first International Conference on the Changing Atmosphere to address the effect of industrial fuel use on atmospheric change. The conference launched the Intergovernmental Panel on Climate Change (IPCC), which today consists of 2,500 scientists. The concern over climate change has continued to grow. In 1992, the Earth Summit was held in Rio de Janeiro, where 132 heads of state approved the Framework Convention on Climate Change to promote an agreement among all nations on how to respond to the growing climate threat—more than 160 countries eventually ratified the convention.

In a 1994 report, the IPCC warned that emissions from the burning of coal and oil were trapping more of the sun's heat than normal. The report cautioned that many serious changes had been noted "including an increase in some regions of the incidence of extreme high temperatures, events, floods, and droughts, with resultant consequences for fires, pest outbreaks and ecosystems."[4] In 1997, the Climate Change Convention was held in Kyoto, Japan, to determine targets and timetables for reducing greenhouse gas emissions.

More than 1,000 scientists worked for two years to produce the recently released report "Climate Change 2001." The IPCC now believes that the earth's temperatures are already rising and will rise by as much as 5.8 degrees Celsius by the end of this century, almost twice the increase predicted in the group's 1995

report. Such an increase will lead to crop failures, water shortages, increased disease, flooding, landslides, and cyclones. The
Global Commons Institute has assessed that damages due to
climate change could amount to $200 billion by 2005 and $400
billion by 2012. By 2050, the property damage could reach $20
trillion. This is why insurance companies are taking climate
change seriously.[5]

The main victims of climate disasters are those who have had
the smallest role in creating climate destabilisation—coastal
communities, small islanders, peasants, and pastoral communities. Small island states, whose very existence can be wiped
off the world map by severe hurricanes, storms, and rising sea
level, have organized themselves into the Alliance of Small Island States (AOSIS) to demand the active reduction of carbon
dioxide emissions by the industrial world. Ambassador A. Tui-
loma Neroni Slade of Samoa captures the spirit of AOSIS: "The
strongest human instinct is not greed ... [I]t is survival and we
will not allow some to barter our homelands, our people, and
our cultures for short-term economic interest."[6]

The AOSIS calls for a 20 percent reduction in 1990 levels of
carbon dioxide emissions by the year 2005.[7] A number of industrial countries advocate similar cuts; Germany and Great
Britain call for a 10 percent cut in emission levels by 2005, and
a 15 percent cut by 2020. The most drastic proposal comes
from Dutch scientists, who call for a 60 to 70 percent carbon
dioxide reduction by the industrial world to stabilize the atmosphere.[8]

Despite the worldwide acknowledgment of climate change
and a commitment to fight global warming, the United States is
a vocal opponent of the Kyoto agreement to reduce greenhouse
gases. When George W. Bush became president of the United
States in 2001, one of his first decisions was to abandon the
agreement and to reverse the US pledge to cut carbon dioxide
emissions from power plants. Bush argued: "Our economy has
slowed. We also have an energy crisis, and the idea of placing caps

on carbon dioxide does not make economic sense."[9] The United States, which produces 25 percent of the world's greenhouse gases, more than any other nation, has officially announced that it will make no cutbacks. Ironically, the United States itself is under serious threat by global warming. Rising sea levels could obliterate the East Coast, as well as the Gulf Coast states of Florida, Alabama, Mississippi, Louisiana, and Texas. The Environmental Protection Agency (EPA) has estimated that a two-foot rise in sea level—caused by increased ocean temperatures and melting ice caps—would wipe out 17 to 43 percent of American wetlands. Total economic losses in North America from weather-related events were $253 billion in the period between 1985 and 1999. The estimated value of incurred coastal property loss as of 1993 was $3.15 trillion.[10] The Midwest also faces a threat from droughts.

Orissa's Supercyclone: A Man-made Disaster

The term *cyclone* is derived form the Greek word *kukloma*, which means the coil of a snake. When fully developed, a cyclone is a vast whirlwind of extraordinary violence, moving at a rate of 300 to 500 kilometers a day along the surface of the sea. When the storm approaches a coastline, the sea level rises rather suddenly and inundates surrounding areas. When the sudden rise in the sea, called a storm tide, strikes, it can devastate an area in a matter of minutes, as was the case with the Orissa super cyclone.

The 1999 cyclone was not a mere natural disaster—it was mainly a man-made ecological crisis unleashed by the combined impact of climate change, industrialization, and deforestation. Climate change is creating climatic extremes in the region. The average wind speed of past cyclones was 73 kilometers per hour; the speed recorded in 1999 was 260 kilometers per hour.[11]

The IPCC surmises that climate change is caused by increasing amounts of anthropogenic greenhouse gases, emitted largely by industrial and corporate activities. These gases increase the tropical sea surface temperatures and intensify tropical rainfall. Such climatic changes and the consequent sea-level rise can have adverse effects on cyclone activity. The rising sea level threatens to inundate lowlands, destroy coastal marshes and swamps, erode shorelines, cause coastal flooding, and increase the salinity of water sources. The worldwide sea-level rise over the next 100 years is expected to most severely devastate the lowlands of the Bay of Bengal. These regions, created by sediment washing down the Ganges, Brahmaputra, and Meghna Rivers, are most vulnerable to submersion. The frequency of these disasters is also expected to increase. One of the necessary conditions for tropical cyclone formation is a sea surface temperature of 26 to 27 degrees Celsius.[12] Global warming is expected to raise the sea temperature and thereby increase the frequency of cyclones.

Destruction of Mangroves

Coastal ecosystems like those in Orissa have mangroves, which, along with shelterbelts, reduce wind velocities and floods. Mangroves absorb the energy of wave and tidal surges, protecting the land behind. The trees also form a barrier against wind. Destruction of coastal mangroves in Orissa, however, has reduced the buffer capacity of coastal ecosystems and allowed storm surges and cyclonic winds to wreak havoc on the region.

Mangroves are also useful in treating effluent, as the plants absorb excess nutrients such as nitrates and phosphates, thereby preventing contamination of shore waters. In regions where these coastal fringe forests have been cleared, tremendous problems of erosion and siltation—and sometimes enormous loss of human life and property—have occurred. Mangrove forests can survive in the saline wetlands because of special features such as their aerial and salt-filtering roots and salt-excreting leaves. Local communities depend on mangrove ecosystems for food, medicine, fuel wood, and construction materials. For millions of indigenous coastal residents around the world, mangrove forests offer dependable livelihoods and sustain their cultures. According to the local communities and forest department in Orissa, the mangroves in the region provide 10 major timber species.

Trade liberalization is one of the leading reasons why mangroves are vanishing. The pressures of trade liberalization and the promotion of export-driven production are promoting shrimp farming throughout the coastal regions. Significant mangrove losses due to aquaculture are especially visible along the west coast of India and in the districts of Karwar and Jumta in Karnataka state, Palghar and Shrivardhan in Maharashtra, and Valsad I in Gujarat. Issukapalli mangrove forests, which once stretched 500 hectares in Andhra Pradesh, have been reduced significantly. All over India, where mangrove forests once stood, roads and aquaculture ponds now lie.

Mangrove forests are desirable for shrimp growth, as they provide important nutrients. In the states of Orissa and West Bengal, numerous shrimp farms have been established in mangrove forests. In the Sunderbans of Bengal, shrimp ponds have been constructed on 35,000 hectares of land once inhabited by mangrove forests. In 1995, the government of Orissa invited project proposals for setting up aqua-farms. This initiative led to an unregulated expansion of aquaculture at the cost of social and ecological sustainability.

The spread of aquaculture in the coastal areas has decreased the coastal zone buffer capacity and has left the regions vulnerable to cyclones and floods and new scales of environmental disaster. In 1991, a tidal wave claimed thousands of lives in Bangladesh as a result of shrimp ponds. A similar wave in 1960 did not harm the villages, as mangroves protected the land at that time. Experts suggest that the destruction caused by the super-cyclone in Orissa could have been minimized had the mangroves along the coastline not been destroyed for shrimp farming: "the (Orissa) coastline was once covered by mangrove forests and these would have dissipated the incoming wave energy"[13] Mangroves export organic matter, providing nutrients to adjacent estuarine and marine ecosystems. The mangrove swamps form the base of the food chain in the sea and coastal waters. The richness of organic matter allows a number of species, both marine and freshwater, to flourish.

Floods and Hurricanes

The supercyclone in Orissa was not an isolated disaster. In the past five years alone, we have heard of hundreds of climate change-related calamities. In 1995, a flood in Bangladesh killed more than 70 people and affected nearly 10 million. In 1995, St. Thomas Island in the Caribbean was reduced to shambles by hurricanes; that same year, France and the Netherlands faced unprecedented rainfall and extensive flooding.

In 1996, the worst cyclone of the century killed 2,000 residents in Andhra Pradesh, India. That same year, a typhoon outbreak in Angola killed more than 600 people. Floods in North Korea led to food shortages for five million people. In March 1996, a deadly blizzard in the western Chinese Highlands pushed at least 60,000 Tibetan herders in Qinghai province and Tibet to starvation by wiping out 750,000 heads of livestock and drastically reducing their food supplies; 48 herders died. The snowfall was four times greater than average, and temperatures fell to −49 degrees Celsius. Also that month, 20 of Laos's rice fields were damaged due to flooding, putting 10 million Laotians at risk of starvation. In June of that year, more than 330 people died in Yemen due to the worst floods in 40 years. The floods caused $1 billion dollars in damages. The stagnant water led to a malaria outbreak, infecting 168,000 people and killing 30.

In 1997, more than 30 people were killed and 120,000 left homeless in the Philippines due to a torrential storm. A succession of ice and rainstorms in the Pacific Northwest caused $25 million of damage that same year. In March, 100,000 farms were wiped out in Bolivia by flooding. That year, 57 people were killed, and thousands in Indiana, Kentucky, Ohio, and West Virginia had to flee their homes, when the Ohio River rose 12 feet above flood levels. The flooding of the Bed River caused $2 billion in damages in Manitoba, Canada, North and South Dakota, and parts of Minnesota.

In January 1998, Peru received 13 liters per square meter within 14 hours. Close to 60 bridges collapsed and 530 miles of highway were wiped out in subsequent weeks. In February, 3,084 people were infected by cholera in Ecuador; 108 died in floods and landslides, and 28,000 lost their homes. In the same year, the Juba and Shabeele Rivers in the horn of Africa flooded, killing 2,000 people and millions of livestock.

Drought, Heat Waves, and Melting Glaciers

While climate change is creating more floods and cyclones, it is also aggravating drought and heat waves. There is either too much water or too little, and both extremes pose a threat to survival. The most dramatic impact of global warming is the melting of ice caps and glaciers. Although there have always been changes in climate, the scientific community and most governments agree that the present crisis of melting glaciers and polar ice caps is ecologically connected to the fossil fuel economy and atmospheric pollution. Snow cover in the northern hemisphere has been reduced by about 10 percent over the past three decades.[14]

Due to climate change, the earth has warmed by somewhere between 0.4 and 0.8 degree Celsius over the past century. The 12 hottest years during the past hundred years have all occurred since 1983, and the three hottest were in the 1990s. Since 1980, average annual temperature has climbed by as much as four degrees Celsius in Alaska and Siberia. In parts of Canada, ice caps are forming two weeks later than they used to, and they break up earlier than in past years.[15]

The rising temperatures are also leading to the melting of glaciers and ice sheets. According to John Michael Wallace, professor of atmospheric science at the University of Washington, "Permanent summertime melting of the whole Arctic could happen in a few decades if trends of the last twenty years continue."[16]

During the past 40 years, there has been a 40 percent decrease in the thickness of perennial Arctic sea ice. Between 1950 and 1970, the ice boundary of the Antarctic Sea shrank by 2.8 degrees of latitude. The annual melt season has increased by up to three weeks in the past 20 years. Between 1961 and 1997, mountain glaciers have reduced by 400 cubic kilometers. The heat being accumulated through the greenhouse effect is accounting for 8,000 joules in terms of the melting of Antarctic and Greenland ice, and 1,100 joules in terms of the melting of mountain glaciers.[17] The Intergovernmental Panel on Climate Change predicts an average increase of global temperatures by 1.5 degrees to six degrees Celsius by 2100.

Glaciers are disappearing in the Alps, in Alaska, and in Washington State. Mount Kilimanjaro, the highest mountain in Africa, has lost 75 percent of its ice cap since 1912. All of its ice could vanish within 15 years.[18] Only two of the six glaciers in Venezuela remain, and if glacial retreat continues at current rates, Montana is expected to lose all the glaciers in Glacier National Park by 2070.[19] According to local people, the Gangotri glacier, the main source of perennial flows of the mighty river Ganges, is receding at five meters a year.[20] The retreat of all glaciers outside the polar region is expected to have contributed to a sea-level rise of two to five centimeters.[21]

The year 1995 was an especially active one: Cadiz in southern Spain, an area that once received the most rainfall in the country, suffered its fourth consecutive year of drought. The rainfall had dropped from 84 inches a year to 37 inches a year. In June, temperatures in Russia reached 93 degrees Fahrenheit, melting the asphalt on roads and airport runways. Northern India also experienced soaring temperatures of 113 degrees Fahrenheit. The heat wave killed 300 people, Around the same time, a heat wave in Chicago killed about 500 people, and Great Britain suffered its hottest summer since 1659 and its driest season since 1721. Northeastern Brazil suffered its worst drought in the century, with rainfall declining by 60 percent. In June 1995, Canadian fires destroyed forests, spreading over 240,000 acres a day. Uncontrollable forest fires also destroyed 700,000 square acres of forestand rangeland in Mongolia.

The calamities were not limited to 1995. In 1996, the worst drought of the century in the United States hit Kansas and Oklahoma, destroying millions of acres of wheat. The United States' wheat reserves dropped to their lowest level in 50 years. In India, consecutive droughts also created food and water crises in Gujarat, Rajasthan, Madhya Pradesh, Orissa, and Chattisgarh. While campaigning for re-election in drought-affected Gujarat in 1999, India's home minister, L. K. Advani, was greeted by people shouting "*Pehle paanni, pbir Advani*" ("First water, then Advani"). In 1997, Rio de Janeiro winter temperatures rose to 108 degrees Fahrenheit. In 1998 more than 13,000 fires spread across Mexico; people were killed, airports were closed, and Mexico City was placed under an environmental alert. As the blanket of smoke moved to the gulf, Texas was put on a health alert.

In September 1997, due to fires in Indonesia and Malaysia, smoke pollution created an emergency. Schools and airports were closed. Ship collisions in the Strait of Malacca killed 29 people, and the haze from forest fires was responsible for an airline crash that claimed 234 lives. Traffic accidents due to low visibility killed hundreds more.

It is the poorest people in the Third World who will be most severely affected by climate change, drought, melting glaciers, and rising sea levels. The peasants, pastoralists, and coastal communities will become environmental refugees as rains disappear, crops collapse, and rivers go dry. The flood risk to coastal communities due to climate change is high: "In extreme circumstances, sea level rise and its associated consequences could trigger abandonment and significant 'off island migration' at great economic and social costs."[22]

Whether water is life-threatening or life-sustaining depends to a large extent on the ability of climate justice movements to end atmospheric pollution and to get rogue countries and rogue corporations to act within the limits of ecological responsibility.

1. Aubrey Meyer, *Contraction and Convergence: The Global Solution to Climate Change* (Totnes, Devon: Green Books for the Schumacher Society, 2000), p. 22.
2. Paul Brown, *Global Warming. Can Civilisation Survive?* (London: Blandford Press, 1996), p, 57.
3. Ibid.
4. Intergovernmental Panel on Climate Change, *Climate Change*, 2001, (Cambridge: Cambridge University Press), p. 1.
5. Meyer, *Contraction and Convergence.*
6. Ross Gelbspan, *The Heat Is On: The Climate Crisis, the Cover-up, the Prescription* (Boulder, CO; Perseus Books, 1998), p, 109.
7. Ibid.
8. "Global Warming Much Worse than Predicted," *The Independent*, June 12, 2001.
9. Jeffrey Kluger, "A Climate of Despair," *Time Magazine*, (April 9, 2001): p. 34.
10. Intergovernmental Panel on Climate Change, *Climate Change*, 2001, p. 363.
11. Vandana Shiva and Ashok Emani, *Climate Change, Deforestation, and the Orissa Supercyclone* (New Delhi: Research Foundation for Science, Technology and Ecology, 2000), p. 4.
12. Ali and Chowdhary, April 1997.
13. Shiva and Emani, *Climate Change, Deforestation, and the Orissa Super Cyclone*, p. 10.
14. Ibid., pp. 810-815.
15. "The Big Meltdown," *Time Magazine*, September 4, 2000, p. 55,
16. John Michael Wallace, *International Herald Tribune*, April 19, 2001.
17. Sydney Levitus, *New York Times*, April 13, 2001.
18. "Climate Crisis," *The Ecologist*, 29: 2.
19. Intergovernmental Panel on Climate Ch ange, *Climate Change* 2001, p. 700.
20. K. S. Foma, *The Traveller's Guide to Uttarakhand* (Chatnoli, India; Garuda Books, 1998), p. 51.
21. Brown, *Global Warming*, p. 87.
22. Intergovernmental Panel on Climate Change, *Climate Change 2001*, p. 856.

CHAPTER THREE

The Colonization of Rivers: Dams and Water Wars

Public Costs and Private Gain: Dams in the American West

Water ownership did not always entail state and private involvement. For a long time, water was under community control. Throughout the world, complex water-conservation and water-sharing systems ensured sustainability and accessibility to all. Community control meant that water was managed locally and as a common resource. Such community-based systems can still be found in the Andes, Mexico, Africa, and Asia.

Community control was eroded when states took control over water resources. In the American West, the state collaborated with private entrepreneurs to acquire water rights. In the Third World, government control was facilitated by giant water-project loans from the World Bank. Dams were a particularly popular means of shifting water control from communities to central governments and colonizing rivers and people. For European colonizers who came to America, river colonization was a cultural obsession and an imperial imperative. Nature in general, and rivers in particular, were valued for their commercial benefit and were seen as being in need of taming. John Widtsoe, an irrigation scientist with the Bureau of Reclamation, once argued:

> The destiny of man is to possess the whole earth; and the destiny
> of the earth is to be subject to man. There can be no full conquest
> of the earth, and no real satisfaction to humanity, if large por
> tions of the earth remain beyond his highest control. Only as all
> parts of the earth are developed according to the best existing
> knowledge, and brought under human control, can man be said
> to possess the earth. The United States ... might accommodate
> its present population within its humid region, but it would not
> then be the great nation that it now is.[1]

W. J. McGee, President Theodore Roosevelt's chief adviser on water programs, projected that the control of water was "the single step remaining to be taken before Man becomes master over Nature."[2] In 1944, describing the blocking of the Sacramento River to build the Shasta dam, the chief of construction, Francis Crove proclaimed: "We had the river licked. Pinned down, shoulders right on the map. Hell, that's what we came up here for."[3]

Rivers following their ecological path were viewed as wasteful: "It would outrage one's sense of justice if that broad stream were to roll down to the ocean in mere idle majesty and beauty."[4] So wrote Wesley Powell, the director of the United States Geological Survey from 1881 to 1899. He also wrote that rivers were "wasting into the sea."[5] President Roosevelt, who founded the Bureau of Reclamation in 1902, shared similar views about water waste. While advocating for the establishment of the Bureau, Roosevelt argued, "If we could save the waters running now to waste, the western part of the country could sustain a population greater than even the legendary Major Powell dreamed."[6]

Although the notion of taming nature justified the construction of massive dams, the limits set by nature did not go unnoticed even by Wesley Powell; it was he who warned against indiscriminate dryland settlement, saying, "It would be almost a criminal act to go on as we are doing now and allow thousands and hundreds of thousands of people to establish homes where they cannot maintain themselves."[7] As early as 1878, Powell had acknowledged the limits to making the desert bloom, and talked of possible dangers for years to come: "I wish to make it clear to you, there is not sufficient water to irrigate all the lands which could be irrigated, and only a small portion can be irrigated," he advised in 1893. "I tell you, gentlemen, you are piling up a heritage of conflict."[8]

By the late 1890s, Los Angeles had already tapped its local supplies, and city officials were secretly purchasing land and water rights in neighboring Owens Valley.[9] In 1907, bonds were issued to finance a 238-mile aqueduct that would divert the eastern runoff of the Sierra Madre. This clandestine agreement to transfer water from the farms to the city led to intense conflict between Owens Valley residents and Los Angeles water users.[10] Nonlocal residents were equipped with private and public investment and backed by the might of the army. In 1924, Owens Valley residents blasted an aqueduct to prevent water diversion to Los Angeles.[11] The water war had begun.

After 12 more blasts, armed guards were stationed on the aqueduct with orders to kill. In 1926, the Saint Francis Dam was built, but it broke soon after, killing 400 people. During the drought of 1929, groundwater pumping began but quickly dried up the 75-square mile Owens Lake. New scarcity had bred new conflicts. In 1976, the aqueduct was bombed again.[12]

Irrigation in the western United States was spurred by the need to provide food for gold-rush miners. By 1890, 3.7 million acres of land were irrigated. But by 1900, many water companies were facing bankruptcy, and public agencies were providing support to private developers.[13] Water projects continued to be driven by the private sector but financed by public investments.

The Hoover Dam on the Colorado River was commissioned by the Bureau of Reclamation during the Great Depression and was completed in 1935. The 726-foot-high dam used 66 million tons of concrete—enough to build a 16-foot-wide highway from New York to San Francisco. The reservoir, Lake Mead, could hold the river's entire flow for two years.

The dam marked the beginning of the large dam era and the partnership between government and corporations in control over water. Six companies—Henry Kaiser, Bechtel, Morrison-Knudson, Utah Construction, MacDonald Kahn, J. F. Shea, and Pacific Bridge—were awarded the bid for the dam. The Colorado River Compact, which approved the dam, excluded local governments and communities from the negotiations and decisions. Native Americans, who had been living in the Colorado River basin for centuries, were completely shut out of the decision to dam the river. As historian Donald Worster observes, "No one asked [Native Americans] to participate in the Colorado Compact negotiations, and the Bureau of Indian Affairs, supposedly their guardian angel, failed to look out for their interests there."[14] Arizona, which considered the dam a theft of the state's natural resources, refused to ratify the compact.

To this day, the primary beneficiary of the Hoover Dam has been California. In fact, the state leads the world in water consumption.[15] Water from the Hoover Dam is transferred to California through a 242-mile aqueduct from the Colorado River, and nearly a third of the hydropower generated by the dam is used to pump water to the state. Although it accounts for a mere 1.6 percent of the 243,000-square-mile Colorado basin, California uses one-fourth of its water. Much of this goes to big farms.[16]

Large water-diversion projects are said to augment water. In reality, they take water from one community to another and from one ecosystem to another. The expansion of irrigated agriculture in the arid American West has come at the cost of agriculture in the eastern and southern parts of the country. Although cotton cultivation on lands irrigated by the Bureau of Reclamation increased by 300 percent in the West, it dropped by 30 percent in the South.[17] In the North, fruit and nut cultivation declined by 50 percent, while it grew by 237 percent in the West; land devoted to bran cultivation fell by 449,000 acres across the United States, but doubled in the West; rice cultivation was abandoned in wet Louisiana while it expanded in the arid West.[18]

Dam construction in the United States was undertaken mainly by the Army Corps of Engineers. Established in 1775, the US Army Corps was once the largest engineering organization in the world. In 1981, the Corps' civil works division alone employed 32,000 civilians and 300 officers, who built over 4,000 civil works, including 538 dams. Today, the Corps operates 150 projects that supply water to industries and urban centers.

The Corps's damming activities extend beyond US borders. During the Green Revolution, dams imposed on the Third World through loan conditions were built mainly by the Army Corps. In 1965, despite a severe drought, the United States government refused to supply wheat to India unless the country altered its policies to introduce irrigation-intensive agriculture.[19]

The task of dam construction was of course assigned to the Army Corps. Loan terms imposed by the United States and the World Bank opened up a worldwide market for dam building. In 1966, President Lyndon Johnson, who had forced India to adopt the Green Revolution, launched a "Water for Peace" program, which called for the Army Corps to build dams in the Third World. In a 1966 speech, he proclaimed:

> We are in a race for disaster. Either the world's water needs will be met, or the inevitable result will be mass starvation.... If we fail, I can assure you today that not even America's unprecedented military might will be able to preserve the peace for long.[20]

Peace and food, the justifications for monumental dam construction, left a legacy of centralized water control, violence, hunger, and thirst. Although the rationale of peace and food emerged 30 years ago, they are still used to justify the control over water by the giant corporations that have replaced the Army Corps.

The Temples of Modern India

Punjab literally means the land of five rivers. The prosperity of the region is intimately linked with the sustainable use of the waters of the Indus and its tributaries, the Jhelum, the Chenab, the Ravi, the Beas, and the Sutlej. Irrigation in Punjab predates the Green Revolution by centuries.

During the ancient time of Greek rule, a flourishing agriculture existed in India, and as far back as the eighth century AD, Arab conquerors differentiated between irrigated and non-irrigated lands for the purpose of levying taxes.[21] Inundation canals and water channels irrigated million of hectares of land. That waterlogging did not occur in these canals was a great advantage. The canals flowed for four to five months during the monsoon season and were dry and served as drainage channels for the remaining part of the year.

The Bhakra Dam was conceived in 1908 with a 395-foot-high-reservoir. In 1927 the height was revised to 1,600 feet. After independence in 1947, the Bhakra Dam assumed a new significance; a large amount of irrigated land in the Indus basin had come under Pakistan's control, so India needed new sources of irrigation for Punjab. The dam was completed in 1963.

Prime Minister Jawaharlal Nehru referred to the Bhakra Dam as the "Temple of Modern India" and used it to shift control over water from the regions and states to the central government. Writing to the Minister for Works, Mines, and Power in 1948, Nehru argued for increased involvement by the central government:

> The Bhakra Scheme is a big scheme and an urgent one, even more urgent than others. Thus far it has been carried on in a spasmodic way, and what surprises me is that the Center has little to do with it although we supply the entire finances. This is entirely unsatisfactory, and I think we should make it clear that we cannot finance a scheme unless we have an effective voice in it. The East Punjab Government has to shoulder tremendous

burdens and in the nature of things they cannot function as effectively as the Center can.[22]

The older canal systems of Punjab were regionally managed within the state. A special unit of the Public Works Department's Irrigation Branch known as the Derajot Circle was established in the 19th century to maintain the inundation canals. After the opening of the Bhakra system, water control was centralized and the Bhakra Beas Management Board established.[23] The centralization of the management system made the Indus basin more vulnerable to floods and led to water scarcity. Water conflicts among neighboring states and between the states and the central government were ongoing.

Nehru, who had once raised the status of dams to temples, confessed later on in his life that he had been victim of the "disease of gigantism." In hindsight, he doubted that the government should have initiated a large dam project like Bhakra because of its cost, its involvement of a considerable amount of foreign exchange, and its lengthy duration. In 1978, Irrigation Minister K. L. Rao made an astute observation about the intrinsic injustice of large dams—that those who bear the cost get no benefits:

When the Bhakra Dam was built, the village of Bhakra, situated on the banks of the Sudej, was submerged and the people built their houses on the adjacent hills. The project resulted in great suffering to the people of the village, but nobody took note of the people's representations. It was many years later, during one of my visits to the dam site, that I found that the new village of Bhakra had neither drinking water nor electricity, though surrounded by blazing brilliant lights. This was indeed unfair, and I asked the Bhakra Management Board to supply both power and water to the village. Even then, there were objections. The management board thought that this was not a proper charge on the project. This indeed was an absurd approach.[24]

In May 1984, the Bhakra Main Canal near Ropar was ruptured. Haryana state incurred a loss of $41,614,648 and saw the damage as an act of sabotage. The governor asked the central government to protect the entire canal in the Punjab territory. The break created a serious water crisis in the state. The breached Bhakra mainline canal, the lifeline for the Sirsa, Jind, and Fatehabad districts of Haryana, forced the government to provide emergency supplies of drinking water by tankers.[25]

In 1986, Prime Minister Rajiv Gandhi reported:

> The situation today is that since 1951, 246 big surface irrigation projects have been initiated. Only 66 out of these have been completed; 181 are still under construction. Perhaps, we can safely say that almost no benefit has come to the people from these projects. For 16 years, we have poured out money. The people have got nothing back, no irrigation, no water, no increase in production, no help in their daily life.[26]

In September 1988, floods drowned Punjab, and 65 percent of its 12,000 villages were marooned. The state suffered a loss of about Rs 1,000 crores and 80 percent of the standing crops were destroyed. Close to 3.4 million people in 10 of the state's districts were affected, and 1,500 people were reported killed.[27]

Experts at Punjab Agricultural University hold that these deaths, and the floods, "were very much man-made with a major share of the blame due to BBMB, the 'Bhakra Beas Management Board.'"[28] BBMB authorities had filled the Bhakra Dam up to 1,687.47 feet, 2.5 feet above the maximum storage capacity, largely for the prime minister's visit for the Bhakra silver jubilee day.[29] The dam released 380,000 cubic meters per second of water into the Sutlej River, already carrying 200,000 cubic meters per second over its capacity of 300,000. Water was similarly released without warning from the Pong Dam. Experts at Punjab Agricultural University maintain:

> The deluge in these areas was not entirely due to rains, as was being made out, but due to criminal water management by the BBMB, who went about irrationally releasing water discharges in lakhs of cusecs without any warning to the thousands of people who live close to embankments of the two rivers.[30]

In November 1988, the chairman of the Bhakra Beas Management Board was fatally shot outside his residence. The floods had aggravated the conflict between Punjab and the central government since BBMB was under the latter's control. In 1986, 598 people were killed in violent conflicts in Punjab; in 1987 1,544 died, and by 1988 the number of deaths had reached 3,000.[31]

Large Dams and Water Conflicts

In the past five decades, the capacity to divert rivers from their natural course increased dramatically with the adoption of technology from the United States. The Bureau of Reclamation and the Army Corps of Engineers were in competition with each other and created a new culture of gigantic constructions financed by public money. Marc Reisner, author of the best-selling book *Cadillac Desert: The American West and Its Disappearing Water*, observes that "What had begun as an emergency program to put the country back to work, to restore its sense of self-worth, to settle the refugees of the Dust Bowl, grew into a nature-wrecking, money-eating monster that our leaders lacked the courage or ability to stop."[32] Interest groups whose positions largely conflicted with those of indigenous populations and ecologists flourished. When the technology euphoria of dam building arrived in India, so did the ecological disruption and social conflict associated with it. These conflicts were magnified because India is a riparian civilization, with settlements and irrigated agriculture taking place along rivers. The regions of India are described by their relationship to rivers or *ab*. Doab is "the land between the Ganges and Yamuna," and Punjab is "the land of five rivers."

The water management systems of both surface and groundwater in the arid and semiarid regions of the Krishna basin have evolved into one of the most sophisticated waterworks in the world. An aerial view of the basin reveals a network of numerous tanks constructed by the local people over an extended period of time. These tanks allow the use of surface water to irrigate approximately 500 acres of land; at the same time recharging the groundwater. They also prevent the easy drainage of water and, therefore, conserve it.

For a long time, these decentralized water conservation systems met both the drinking and the agricultural water needs of the surrounding communities. There was no major long-distance transfer of water, and local cropping patterns evolved in accordance with the local water endowment.

The needs of the Vijayanagar Empire led to the first major intervention in the natural water flow. During the reign of King Krishnadevaraya in the 16th century, there were many attempts to divert the water of Tungabhadra. The rulers of Vijayanagar, who understood the crucial role of tanks in food production and drinking water provision, undertook a systematic program of tank construction.[33] The Daroji and the Vyasayaraya Samudram tanks in Cuddapah district are the result of this program. Although the Vijayanagar irrigation systems diverted water to some extent, they never caused waterlogging because they functioned as "round rivers," diverting the water from the river but carrying it back through drainage channels. By contrast, large dams built on the same river led to immediate waterlogging.[34]

Dams and Displacement: The Case of India

River valley projects are usually considered the solution for agricultural water needs, flood control, and drought mitigation. In the past three decades, India has seen the erection of some 1,554 large dams. Between 1951 and 1980, the government spent $1.5 billion on large or medium irrigation dams. Yet the return from this large investment has been far lower than anticipated. Where irrigated lands should have yielded at least five tons of grain per hectare, output has remained at 1.27 tons per hectare.[35] The annual loss due to unexpected low water availability, heavy siltation, reduced storage capacity, and waterlogging now amounts to $89 million.[36]

The Kabini project in Karnataka is a perfect illustration of how water development projects can themselves disrupt the hydrological cycle and destroy water resources in basins. While the dam submerged 6,000 acres of land, relocating displaced villages required the clearing of 30 thousand acres of primeval forests.[37] Local rainfall fell from 60 inches to 45 inches, and high siltation drastically reduced the life of the dam. Within two years, waterlogging and salinity destroyed large areas of coconut and paddy fields nearby.[38]

The damming of two of India's most sacred rivers, the Ganges and the Narmada, has generated vehement protests from women, peasants, and tribals whose life-support systems have been disrupted and whose sacred sites have been threatened. The people of Narmada Valley are not merely resisting displacement due to the Sardar Sarovar and Narmada Sagar Dams; they are waging war against the destruction of entire civilizations. As the internationally acclaimed novelist Arundhati Roy puts it:

> Big dams are to a Nation's 'Development' what Nuclear Bombs are to its Military Arsenal. They're both weapons of mass destruction. They're both weapons Governments use to control their own people. Both Twentieth Century emblems that mark a point in time when human intelligence has outstripped its own instinct for survival. They're both malignant indications of civilization turning upon itself. They represent the severing of the link, not just the link—the understanding—between human beings and the planet they live on. They scramble the intelligence that connects eggs to hens, milk to cows, food to forests, water to rivers, air to life and the earth to human existence.[39]

Over the past two decades, many men and women have dedicated their lives to the protest of the damming of Narmada Valley and the Ganges. Since the 1980s, two old men have been engaged in satyagraha (Gandhian nonviolence)[40] on the banks of the two rivers. Sunderlal Bahuguna has been living in a small hut at the Tehri Dam site on the Ganges to block the flooding of Tehri and stop the building of a dam on an earthquake fault. Baba Amte, who resisted dam building in Maharashtra, has been stationed on the banks of the Narmada for years. In 1984, Amte wrote a letter to the prime minister, in which he referred to the dams as genocide.[41] Although bedridden due to a severe back problem, he still remains by the valley and says he will go with the river. Medha Patkar, a leading activist of the Narmada Bachao Andolan, and Arundhati Roy have also committed themselves to the fight against the Narmada Dam project—the world's largest water project.

The Narmada project consists of 30 large, 135 medium, and 3,000 minor dams on the Narmada River and its tributaries. It is expected to uproot one million people, submerge 350,000 hectares of forest, drown 200,000 hectares of cultivable lands, and cost $52.2 billion over the next twenty-five years.[42] The Sardar Sarovar Dam, already under construction, is facing major opposition from human rights and environmental groups as well as tribals likely to be displaced. The dam threatens people in 234 villages.[43] Next in line for contruction is the Narmada Sagar project, which promises to submerge 91,348 hectares of land and displace people from 254 villages.[44]

The Narmada Valley protest, which once was a fight for a just settlement of the displaced people, has rapidly evolved into a major environmental controversy, calling into question not only the method of compensation for the evictees but the logic of large dams altogether. The movement has taken inspiration from earlier successful struggles that led to the withdrawal of two major dam proposals—the Silent Valley and the Bodhghat Dam projects. Large coalitions of local communities, environmentalists, and scientists worked together in the 1980s to stop these dams. As dam tensions emerge and grow, they will not only address problems created upstream due to submergence, they will also raise questions about problems created downstream due to water overuse and misuse by intensive irrigation.

The construction of the Ukai Dam across the Tapi River in Gujarat displaced 52,000 people.[45] The farmers who once occupied fertile agricultural lands were forced to resettle in an area cleared of forests. Prior to their settlement in their new site, the government promised to level the land, clear the tree stumps, sink wells free of charge, and install power connections.

Once the farmers arrived, however, they found that most of the promises were not kept. The land was leveled with some assistance from the government, but the farmers cleared the tree stumps themselves with great difficulty. Moreover, the clearing of the forest and the removal of the remaining tree stumps led to topsoil erosion and made farming impossible. The government reneged on the wells, saying it had promised to sink wells only for those who had wells in the old villages. However, most of the old villages were near the river, and not many farmers had needed to sink wells. With insufficient water, little food, and almost no work, the settlers soon became migrant laborers in surrounding sugarcane fields.The Pong Dam in Himachal Pradesh in the Himalayas displaced 16,000 families. The government then attempted to rehabilitate about half of them in the faraway deserts of Rajasthan, and each family was given 16 acres of land—the largest compensation in the country so far. Despite these efforts, the families were unable to adjust to the new climate, water, people, and language, and most of them sold their lands and returned to their native place.

The Bhakra Dam was responsible for the displacement of 2,180 families of Bilaspur in Himachal Pradesh.[46] The families, who were promised land in surrounding Haryana 25 years ago, have yet to be fully compensated; only 730 families (33 percent) have been rehabilitated. Moreover, while the land, taken from them between 1942 and 1947 was estimated at the prevailing rates during that time, the lands they actually received were appraised at rates prevailing between 1952 and 1957, leaving them with a mere one to five acres per family. Like the people displaced by the Pong Dam, they too fled their harsh new environment and returned, to Himachal Pradesh.[47]

Dam conflicts in the past revolved around displacement. Today, the ecological imperative for the protection of nature has added a new dimension to the struggle of displaced people. They are now fighting for their own survival as well as for the survival of their forests, rivers, and land. In east India, tribals of 121 villages, who faced eviction by the Koel-Karo project in Bihar, successfully stopped the construction work.[48] Had the project been completed, it would have taken water from the Koel River at Basia and diverted it to another dam near Lohajamir village in Topra block, Ranchi district, and to the Karo River.

It would also have submerged more than 50,000 acres of land, including 25,000 acres of forests under tribal control by customary law.

In postcolonial India, most large dams have been financed by the World Bank. I was personally involved in assessing the impact of World Bank–financed dams on the Krishna, Kallada, Suvernarekha, and Narmada Rivers. In each case, the ecological and social costs far surpassed the benefits. Typically, the benefits were grossly exaggerated in order to accommodate the World Bank's logic of returns on investment.

The Sri Sailam Dam on the Krishna River is among the hundreds of dams financed by the bank. In the summer of 1981, the government evacuated local residents from the area with the assistance of police and bulldozers. The experience in Sri Sailam is illustrative of the hidden cost of building large dams in India. Each water development project leaves behind evictees whose lives are violently overturned.

Costs should never be assessed purely in commercial terms. The Suvernarekha Dam was built with a $127 million loan from the World Bank, primarily to provide industrial water for the expanding steel city of Jamshedpur.[49] The dam displaced 80,000 tribals. In 1982, Ganga Ram Kalundia, the leader of the tribal anti-dam movement, was shot and killed by the police. Even after his death, Kalundia's fellow tribals continued the struggle:

> Our links with our ancestors are the basis of our society and of the reproduction of our society. Our children grow up playing around the stones which mark the burial sites of our ancestors. ... Without relating to our ancestors, our lives lose all meaning. They talk of compensation. How can they compensate us for the loss of the very meaning of our lives if they bury these burial stones under the dam? They talk of rehabilitation. Can they ever rehabilitate the sacred sites they have violated? [50]

The massive people's movements managed to force the World Bank out of the Narmada Valley Dam. But the bank stepped out of one project only to deepen its grip on India's water resources through more loan conditions. World Bank–driven policies of water privatization are shifting control from governments to corporations. The centralization of power over water through development projects makes this transition easier. With communities bypassed, the World Bank and indebted governments are making frantic deals with corporations to own, control, distribute, and sell our scarce water resources.

The Global Picture of Displacement

While large dams in India have displaced between 16 million and 38 million people, in China, 10 million people have been displaced by the Three Gorges Dam in the Yangtze River Valley alone. The World Commission on Dams estimates that worldwide, 40 to 80 million people have been displaced by dam projects.[51] The commission concludes that too often "an unacceptable and often unnecessary price has been paid to secure those benefits, especially in social and environmental terms, by people displaced, by communities downstream, by taxpayers, and by the natural environment."

Worldwide, an estimated $2 trillion has been invested in more than 45,000 large dams. Between 1970 and 1975, the peak period of dam building, nearly 5,000 large dams were built all over the world. The top five dam-building countries account for 80 percent of all large dams, and China, with 22,000, accounts for 50 percent of them.[52] The United States is home to 6,390 large dams, closely followed by India, with 4,000, Japan with 1,200, and Spain with 1,000. While dam construction has slowed down in the United States and Europe, India is experiencing the largest amount of dam construction in the world and accounts for 40 percent of dams currently under way. It is no surprise that the most contentious battles over dam construction are taking place in India.

Displacement is an intrinsic aspect of wars unleashed by large water projects. People fiercely resist being forced out of their homes and losing their livelihoods. Unfortunately, anti-dam movements in the Third World are facing new violence from states acting in partnership with global corporations. The World Commission on Dams reports that during the construction of the Kariba Dam in Africa, resistance by the Tonga people was met with state repression; eight were killed and 30 were injured.[53] The report also notes that in April 1980, police in Nigeria fired at people protesting the Bakolori Dam and in 1985, 376 women and children in Guatemala were murdered to make way for the Chixoy Dam.

In 1991, 16,350 tribal families threatened by the Koel-Karo Dam in India successfully stopped its construction through their movement, the Koel Karo Jan Sangathan. The dam would have displaced residents from 256 villages and submerged 152 sacred ancestral graves. The government has now resorted to force to end the people's resistance, which has lasted more than 10 years. In February 2001, during a demonstration against the assault of one member of the Koel Karo Jan Sangathan, protestors were met with gunshots from the police: six people, including three children, died on the spot and 50 were wounded.[54]

River Diversions and Water Wars

Large dams are constructed to divert water from the natural drainage of rivers. Altering a river flow also modifies the distribution patterns of water in a basin, especially if interbasin transfers are involved. A shift in water allocation most often generates interstate conflicts, which rapidly escalate into disputes between central governments and states.

Every river in India has become a site of major, irreconcilable water conflicts. The Sutlej, Yamuna, Ganges, Narmada, Mahanadi, Krishna, and Kaveri Rivers have been the center of heated court cases among states that disagree over ownership and distribution of water. Even events such as the kidnapping of the popular Indian film star Rajkumar in Karnataka by the forest bandit Veerappan in 2000 was related to the conflict over Kaveri River waters between Karnataka and Tamil Nadu. Veerappan's demands included more of the Kaveri's water for Tamil Nadu.[55]

The Kaveri River is one of the rivers involving an intricate interstate dispute. The Kaveri has been used for centuries and the famous, 2,000-year-old Grand anicut structure on the Kaveri River is believed to be the oldest water-flow management system on the Indian subcontinent. When the British introduced their engineering system in Thanjavur at the Kaveri basin in 1829, they could not handle the siltation and flooding and eventually reverted to the ancient anicut system.

Since India's independence, the Kaveri has became the most contentious river between the states of Tamil Nadu and Karnataka. Water wars between these two states have led to bloodshed and brought down governments.[56] Although much of the recent conflict has stemmed from a decision by the Kaveri Water Disputes Tribunal to reduce Tamil Nadu's water supply from the Kaveri, the dispute can be traced to the 1892 agreement between the Madras state (now Tamil Nadu), which was under British rule and Mysore, which was under indirect colonial rule. In 1892, the British awarded Madras, the lower riparian state, veto power over all irrigation works undertaken by Mysore, the upper riparian state. In 1924, Madras and Mysore came to an agreement to build the Krishnaraj Sagar Dam and irrigate an additional 100,000 acres of land.

In 1974, the irrigation extension agreement between Madras and Mysore, renamed Tamil Nadu and Karnataka respectively after India's independence, expired and the conflict over sharing Kaveri waters resurfaced. In 1983, the dispute reached the Supreme Court when the Tamil Nadu Farmers Society filed a petition for a greater share in the Kaveri waters.[57] The court asked the central government to form the Kaveri Waters Dispute Tribunal in 1990.

However, the interim measures, ordering Karnataka to release water on a weekly basis, could not be implemented. When the tribunal issued its order, Karnataka passed an ordinance to block its implementation. The president of India had to intervene and refer the case back to the Supreme Court in 1991. The court deemed Karnataka's ordinance beyond the legislative competence of the state and upheld the tribunal's decision. The court's verdict triggered riots in Karnataka's capital, Bangalore. Tamilians were attacked and driven off their farms, and their houses were looted and burned. The violence spread to Tamil Nadu, and this time Kannadigas were attacked. The water riots of 1991 displaced an estimated 100,000 people.[58]

In the Americas, conflict between the Unites States and Mexico over Colorado River waters has intensified in recent years. In 1944, a treaty allocated a 1.5 acre-foot of Colorado River water to Mexico. In 1961, Mexico protested that water flowing from the United States was heavily salinated by dams at Glen Canyon, Lake Mojave, and Lake Mavasu as well as the Hoover Dam.[59] In 1974, the United States built a plant to desalinate the Colorado River water before it entered Mexico. The total cost to build the project was $1 billion. Irrigation water provided for

$350 per acre-foot in the United States cost $300 per acre-foot for desalination alone.[60]

Hydro-Jihad

Large dam-related conflicts are not restricted to states—they also involve wars between nations. The Tigris and Euphrates Rivers, the major water bodies sustaining agriculture for thousands of years in Turkey, Syria, and Iraq, have led to several major clashes among the three countries. Both rivers originate in Eastern Anatolia, Turkey, and the country holds absolute sovereignty over water in its territory. Turkey's position is "The water is as much ours as Iraq's oil is Iraq's."[61] On the other hand, to assert its historical rights, Iraq invokes the "prior use" doctrine, which bases water rights on the cowboy logic of "first in time, first in right" and traces the use of the rivers to the people of Mesopotamia 6,000 years ago.[62] In recent years, conflicts have been triggered by increased water demands for industrialization. Turkey created the State Hydraulic Works in 1953 to construct large dams and hydroelectric projects.[63]

The Ataturk Dam is at the center of the Southeast Anatolia Project (GAP).[64] The dam, completed in 1990, transfers water via a 26-kilometer tunnel to the Harran Plain in southern Turkey. The conflict between Iraq and Turkey is expected to intensify as Turkey attempts to move with its $32 billion plan to build 22 dams on the Euphrates for the irrigation of 1.7 million hectares of land.[65] When the two dams operate along with the Ataturk Dam, Iraq would lose 80 to 90 percent of its allotment of Euphrates water.[66]

Water development projects on the Euphrates have been the cause of armed conflict between Turkey, Syria, Iraq, and the Kurds. In 1974, clashes took place between Syria and Iraq. The PKK, the Workers' Party in Turkey, has threatened to blow up the Ataturk Dam and the GAP.

The Kurds, who are divided across Turkey, Syria, and Iraq, have launched nationalist movements in each state. Between

1950 and 1970, more than a million Kurds migrated to the west where the PKK continued to fight domestically. And in 1989, Turgut Ozal, then prime minister of Turkey, threatened to use water against the militants by cutting supplies off entirely unless Syria expelled the PKK, to whom it was giving refuge. In 1998, the Turkish chief of staff announced a "state of undeclared war" with Syria.[67]

Ethnic wars and water are intimately intertwined, as is illustrated in the case of the Ilisu Dam, which is expected to displace 78,000 people in the Kurdish region of southeast Turkey, and destroy the historic town Hasankeyf. The local communities do not want the dam, but their fear of being identified as part of the separatist movement keeps their resistance underground. The Ilisu fact-finding mission reports that "the authorities' association of opposition to Ilisu with separatism is a major deterrent to any meaningful dissent. Put bluntly, people are frightened to take a public position against the dam." The dam is clearly a means of political control. According to a member of the state police, "The dam means power—who has the water has the power."[68]

Although the Middle East is water-scarce, water projects in the region are grandiose. Iraq's river project, the 560-kilometer-long artificial Saddam River, cuts across the Tigris and Euphrates Rivers. The huge diversion scheme has turned 57 percent of former marshland into dryland and now threatens the survival of marsh Arabs who have been living by those rivers for 5,000 years. In defense, the marsh Arabs have declared what they call a "hydro-jihad" on Iraq.[69]

Israel and the West Bank

The war between Israelis and Palestinians is to some extent a war over water. The river under contention is the Jordan River, used by Israel, Jordan, Syria, Lebanon, and the West Bank. Israel's extensive industrial agriculture requires the river's water as well as the groundwater of the West Bank. While only 3 percent of the Jordan basin lies in Israel, the river provides for 60 percent of its water needs.[70]

Israel's very formation was based on ensuring access to water. "It is necessary that the water sources, upon which the future of the Land depends, should not be outside the borders of the future Jewish homeland," wrote Israel's former prime minister David Ben-Gurion in 1973. "For this reason we have always demanded that the Land of Israel include the southern banks of the Litani River, the headwaters of the Jordan, and the Hauran Region from the El Auja spring south of Damascus."[71]

Water conflicts began in 1948, when Israel undertook the National Water Carrier Project, which involved a gigantic water pipeline extending from the Jordan River to the Negev Desert to irrigate crops.[72] This project led to a dispute with Syria. In 1953, United States envoy Eric Johnston initiated the Unified Development of Water Resources plan to resolve conflicts between Israel, Syria, and Jordan. Syria rejected the plan, and since then, Israel-Syria border conflicts have been closely connected to river diversions by Israel. Former Israeli prime minister Levy Eshkol declared in 1962 that "water is the blood in our veins" and that being prevented from accessing it would be cause for war.[73]

Between 1987 and 1988, Israel used 67 percent of its water for agriculture and allocated the rest for domestic and industrial purposes.[74] Although Israel's agricultural water consumption had been reduced to 62 percent by 1992, it remained the leading sector for water use. In 2000, 50 percent of the total cultivated area in Israel was irrigated; in contrast, Palestinian villages consumed only two percent of Israel's water.[75] The water apartheid, demarcated along ethnic and religious lines, is fueling the already heated Israeli-Palestinian conflict.

The 1967 war, which led to the Israeli occupation of the West Bank and the Golan Heights, was in effect an occupation of the freshwater resources from the Golan Heights, the Sea of Galilee, the Jordan River, and the West Bank. As Middle-Eastern scholar Ewan Anderson, notes, "The West Bank has become a critical source of water for Israel, and it could be argued that this consideration outweighs other political and strategic factors."[76]

Between 1967 and 1982, West Bank waters were controlled by the military. Now they are controlled by Israel's water company, Mekorot, and integrated into Israel's overall water network.[77] West Bank waters supply 25 to 40 percent of Israel's water; Israel consumes 82 percent of the West Bank's water, while Palestinians use 18 to 20 percent. Palestinian water use is controlled and restricted by the Israeli government. A 1967 military order decreed:

> No person is allowed to establish or own or administer a water institution (any construction that is used to extract either surface or subterranean water resources or a processing plant) without a new official permit. It is permissible to deny an applicant a permit, revoke or amend a license, without giving any explanation. The appropriate authorities may search and confiscate any water resources for which no permit exists, even if the owner has not been convicted.[78]

In 1999, Palestinians were allowed to dig only seven wells.[79] In addition, Palestinian wells could not exceed 140 meters in depth, while Jewish wells could be as deep as 800 meters.

As drought and overuse aggravate the water scarcity, water conflicts are bound to intensify. The water level of the Sea of Galilee is at a 100-year low; since 1993, it has fallen 13 feet. Because of drought, Israel had to reduce its water use in agriculture by 10 percent in 1999. It is predicted that Israel will have to cut water use further, cease its cultivation of cotton and oranges, and shift to drought-resistant crops.[80]

Conflict over the Nile

The Nile is the longest river in the world and is shared by 10 African countries, including Ethiopia, the Sudan, Egypt, Uganda, Kenya, Tanzania, Burundi, Rwanda, the Democratic Republic of Congo, and Eritrea. It is also another complicated site of water conflict. In 1990, the total population of the Nile basin countries was estimated at 245 million and projected to reach 859 million by 2025. Ethiopia contributes 86 percent of the total annual flow of the Nile, while the remaining 14 percent comes from Kenya, Uganda, Tanzania, Rwanda, the Democratic Republic of Congo, and Burundi.

The White Nile, arising in Burundi, and the Blue Nile, originating in Ethiopia, have led to historical conflicts between Egypt, Ethiopia, and the Sudan. During their colonial rule in the Sudan, the British, who used the Nile in the Sudan for navigation, signed an agreement with Ethiopia in 1903 not to manipulate the flow of the Blue Nile.[81] In 1958, Egypt began building the Aswan Dam and displaced 100,000 Sudanese.[82]

Initially the Aswan Dam generated conflicts between Egypt and Sudan. But the Sudanese were placated with the promise of more water. However, Ethiopia was never consulted in the sharing of Nile waters and retaliated by declaring its right to use the Nile whichever way it chose. Upon the completion of the dam in 1970, Egypt and the Sudan began construction of the Jonglei Canal for $100 million until the Sudanese People's Liberation Army ceased the project and drove out the construction crew.[83]

In 1959, Egypt and the Sudan entered a bilateral agreement known as the "Full Utilization of the Nile Waters," dividing the entire flow of the Nile among themselves, regardless of the water demands, potential or otherwise, of the upper riparian states. This agreement has been a source of unending battle between the three countries.[84] In the 1960s, Emperor Haile Selassie of Ethiopia, through a loan from the African Development Bank, hired the US Bureau of Reclamation to build 29 irrigation and hydroelectric dams on the Blue Nile.[85] However, Egypt, whose water supply stood to be reduced by 8.5 percent by the new dams, blocked the approval of the loan and prevented the projects.

In 1997, the United Nations held the Convention on the Law of the Non-Navigational Uses of International Water Courses in order to create guidelines for water sharing of international rivers. The two principles used at the convention were the rule of equitable and reasonable use and the no-harm rule: equitable use referred to water sharing on an equitable basis among multiple users, and the no-harm rule referred to not causing harm to co-riparian states.

Enforcing these rules gave rise to diverse interpretations and hence conflict. These two rules have been invoked by Ethiopia, Egypt, and the Sudan and have led to more intense debates over water use. On the one hand, Egypt and the Sudan have held that the 1959 Nile agreement is non-negotiable on the basis of the no-harm rule. On the other hand, Ethiopia and other upstream countries have used the principle of equitable use among co-riparian states to argue for their water rights.[86]

In February 1999, at the meeting of the Council of Ministers of Water Affairs of the Nile Basin in Tanzania, the Nile Basin Initiative was launched. The 10 Nile basin states endorsed a Nile River Basin Strategic Action Program with a vision "to achieve sustainable socioeconomic development through the equitable utilization of water resources and has recognized the rights of each riparian state to use the resources of the Nile within its boundaries for development."[87] The countries are trying to go beyond past conflicts and sustainably and justly use the waters of the world's biggest river for some of the world's poorest people.

International Water Rules

Neither international nor national water laws adequately respond to the ecological and political challenges posed by water conflicts. No legal document in contemporary law mentions the most basic law related to water—the natural law of the water cycle. Claims are derived from and protection is limited to artificial concrete structures. This limitation has propelled regions and states to enter a contest for the most extravagant water projects as a means of establishing their rights to water—the more you extract and divert water through giant projects, the more you can claim rights. Water conflicts continue to escalate and, to date, no appropriate legal framework exists to resolve these conflicts.

Four theories of water rights—the territorial sovereignty theory, the natural water flow theory, the equitable apportionment theory, and the community of interest theory—have guided water distribution practices around the world. The territorial sovereignty theory of 1896, also known as the Harmon doctrine, holds that riparian states have exclusive or sovereign rights over the waters flowing through their territory. Countries can use this water any way they choose, regardless of their infringement on other riparian states. This doctrine has been relevant in the dispute between the United States and Mexico over the Rio Grande.

The Harmon doctrine has never won complete acceptance because it violates the concept of justice. Even countries that benefit from the rule have conceded rights to the lower riparian users. While arriving at a settlement with the other riparian states, even the United States, originator of the Harmon rule, has granted some rights on the ground of good-neighbor policy. In a 1906 treaty regarding the Rio Grande, while affirming the Harmon doctrine, the United States was "willing to provide Mexico with water equivalent to that which she had used before the diversions took place" on the basis of "international comity."[88] Again in 1944, a treaty between the two granted Mexico the right to a specified quantity of waterfrom the Colorado River. Similarly, India, while claiming absolute supremacy as the riparian owner of the Indus River, has conceded rights to neighboring Pakistan.[89]

The natural water flow theory, also known as the territorial integrity theory, maintains that since a river is a part of the territory of the state, every lower riparian owner is entitled to the natural flow of the river, unhampered by the upper riparian owners. The upper riparian owner must allow the water to flow in its natural course to the lower riparian owner in its ordinary channel with reasonable use by the upper riparian owner. This principle was derived from British private property laws and applied to water in a unitary state. Egypt used this doctrine in 1952, against the Sudan, claiming absolute water use of the Nile. However, the Nile Waters Commission rejected Egypt's claim. In 1929, Egypt scored a victory when Britain awarded it veto power over utilization of water by upper riparian states.[90]

The theories of equitable utilization and community of interest are closely related. Equitable utilization holds that international rivers should be used by different states on an equitable basis. In recent years, the equitable utilization theory has gained international acceptance. The Helsinki Rules on the Uses of the Waters of International Rivers adopted in 1966[91] recognized that states are "entitled to a reasonable and equitable share in the beneficial uses of waters of an international drainage basin." The rules overturned those of the American West and established that an existing use may have to give way to a new use for equitable distribution.

Although popular, the theory of equitable distribution is not without problems. The most difficult question lies in the meaning of equitable distribution. The equitable apportionment criterion used to resolve interstate conflicts does not lend itself to precise articulation; dividing a river is not an easy task. The underlying principle of equitable apportionment is equitability, not equality. Equitable utility is defined as the maximum benefit accruing to all the riparian states, in light of their differing economic and social needs.

This dual goal of achieving full benefit while catering to varied needs is precisely what poses a challenge: every state and river is unique, and a solution in one case may not be feasible in another. Crafting guidelines for equitably sharing water requires an analysis of complex technical and economic data as well as a judicious balancing of competing claims and uses of the river. The problem is further complicated because water use is normally determined by the needs and economic development stages of the nation—factors that are constantly changing.

Despite the difficulties inherent in the doctrine of equitable utilization, the International Law Association and the United Nations have offered broad guidelines and fundamental principles. According to the Helsinki Rules on the Uses of the Waters of International Rivers, "each basin state is entitled, within its territory, to a reasonable and equitable share in the beneficial uses of the waters of an international drainage basin." The need now is to combine ecology with equity, and sustainability with justice.

During the period of the large dam euphoria, river diversions were assumed to offer only benefits and no costs. However, as we enter the era of ecological constraint, the principle of equitable use, previously defined in purely economic terms, requires a radical alteration in order to preserve the integrity of river basins and minimize water conflicts. Current applications of water rights largely uphold the rights of a state to control or consume water through large water projects. The creation of the Krishna Valley Authority (KVA) in India is an illustration of how the theory of equitable use favors large dam construction.

The Krishna Tribunal established the KVA in order "to ensure that the waters of the river Krishna are stored, appropriated and used to the extent and in the manner provided."[92] The Krishna Valley Authority, modeled after the Tennessee Valley Authority, was not created to conserve and protect Krishna River; its purpose was to engage in integrated planning at the level of the entire basin. As Marc Reisner points out, "The creation of the Tennessee Valley Authority marked the first time a major river system was 'viewed whole' even if the natural river disappeared as a result."[93]

The framework of scientific knowledge and social justice currently used in water conflict resolution assumes that a river is wasted if it is not dammed. The concept of protective use gives priority to dam building and other water project construction. The Helsinki Rules state that an existing reasonable use is acceptable "unless the factors justifying its continuance are outweighed by other factors, leading to the conclusion that it be modified or terminated so as to accommodate a competing incompatible use." If the existing use is held to be conclusive, then "it freezes river development according to the requirements of the earlier user. Indeed, it is conceivable that, if a state moves quickly enough, it could appropriate all of the waters of a basin to the complete exclusion of its co-basin states."[94] But if no weight is given to the existing uses, it would inhibit river development, as no nation would like to invest large sums of money on projects without assured continuation of water use. The Helsinki Rules represent a compromise between the conflicting forces involved in dam building.

In India, no state can have free rein over a common source of water such as an interstate river. The Government of India Act of 1935 imposes limitations on provinces' use of interstate river waters. If the action of one province affects or is likely to prejudicially affect the interest of another province, the latter can file a complaint to the governor-general. The Indian Constitution also bars a co-riparian state from developing an interstate river without taking into account the harm it can cause to other co-riparian states. The Constitution empowers Parliament to provide "for the adjudication of any dispute or complaint with respect to the use, distribution or control of the water of or in any interstate river or river valley."[95] The act, however, is silent on what principles ought to be followed in settling interstate water disputes.

The existence of international guidelines such as the Helsinki Rules and the United Nations Convention on the Law of Non-Navigational Uses of International Water Courses does not necessarily ensure justice. Each basin is so distinct that a monolithic water-use approach would be unfeasible. In light of ecological diversity, the principle of equitable utilization becomes vague. Equitable use theory treats rivers as static resources to be apportioned at will. When it comes to rivers, what is in fact appropriated is the flow; and because water is a flow and not a stock, its distribution has nonlocal impact. The distribution of benefits and losses to upstream and downstream regions or to riparian and nonriparian states, changes over time, as does the implication for equitable sharing.

The issue of water rights allocation is not only one of maintaining a balance between territorial sovereignty and riparian rights; water projects also have a severe ecological impact, and costs are unequally distributed among states and among social groups. While natural flow cannot be an absolute criterion, conservation must be a criterion for determining sustainable use. The ecological perspective also helps correct the view that water conserved is water wasted. Ecologically unexploited water can be critical in maintaining essential ecological processes such as groundwater recharge and freshwater balance.

The ecological links between surface water and groundwater and between freshwater and the life in the ocean have been overlooked in resource management as well as in legal frameworks. In Krishna, groundwater use was disassociated from the utilization of the Krishna waters, and the Krishna Tribunal granted states full freedom to use groundwater. By excluding control over groundwater utilization, the tribunal allowed privatization and overuse of water resources and fostered an environment for new conflicts. Groundwater use was unregulated and thus depleted in almost all parts of the basin, further aggravating water scarcity and drought. The lack of regulation also introduced new demands for river diversions and interbasin transfers.

In the Rayalseema region, overexploitation of groundwater and the collapse of the indigenous system of irrigation have given rise to new demands for the interbasin diversion of the Krishna basin waters. Surface water and groundwater cannot be artificially separated, since surface water flows recharge groundwater, and groundwater depletion affects the status of surface waters.

Disputes over dams are struggles among communities and regions about how much water one region can take from another, or how much environmental damage one group must bear in order that another group can meet its irrigation or energy needs. So far, struggles against dams in India have largely originated from the problem of displacement. It is a struggle between displaced citizens and the ruthless state machinery.

On the other hand, struggles against massive irrigation systems' byproducts, such as waterlogging and salination, are often limited to challenging the distribution of large water projects and have not focused on large-scale storage systems. Both the ecological impacts of storage—submerging forests, homelands, and farmlands—and the impacts of canals and irrigation need to be taken into account. Finally, water rights conflicts have predominantly taken the form of interstate conflicts at the regional level.

A coherent framework for a just and sustainable water-use policy can evolve only when there is a dialogue between the movement against dams, the movement against the ecological hazards of intensive irrigation, and the movement for water rights. The key to linking these movements is the ecological perspective, which connects water to its various functions in river basins. An ecological paradigm allows for an ecological audit of water projects, exposes the hidden costs of such projects, and proposes an alternative for resource allocation.

1. John Widtsoe, "Success on Irrigation Projects" (published as a pamphlet in 1928), p. 138.
2. Charles R. Goldman, James McEvoy III, and Peter J. Richerson, eds., *Environmental Quality and Water Development* (San Francisco: W.H. Freeman, 1973), p. 80.
3. "By a Damsite," *Time Magazine*, June 19, 1994, p. 79.
4. Paul Shepard, *Man in the Landscape: A Historic View of theAsthetics of Nature* (New York: Knopf, 1967), p. 141.
5. Fred Powledge, Water: *The Nature, Uses, and Future of Our Most Precious and Abused Resource* (New York: Farrar, Straus and Giroux, 1982), p. 279.
6. Bureau of Reclamation, "Reclamation" (Washington, DC, 1975).
7. Tim Palmer, *Endangered Rivers and the Conservation Movement* (Berkeley, CA: University of California Press, 1986), p. 20.
8. Ibid.
9. Ibid., p. 22.
10. Donald Worster, *Rivers of Empire: Water, Aridity, and the Growth of the American West* (New York: Pantheon Books, 1985), p. 202.
11. Palmer, *Endangered Rivers*, p. 58.
12. Ibid.
13. Worster, *Rivers of Empire*, p. 98.
14. Ibid, p. 211.
15. Ibid.
16. Ibid.
17. Palmer, *Endangered Rivers*, p. 215.
18. Ibid., p. 183.
19. Vandana Shiva, *Violence of the Green Revolution* (London: Zed Books, 1988).
20. Worster, *Rivers of Empire*, p. 264.
21. Vandana Shiva and Radha Holla Bhar, *History of Food and Farming in India* (New Delhi: Research Foundation for Science, Technology, and Ecology, 2001).
22. Shiva, *Violence of the Green Revolution*.
23. Ibid.
24. L. C. Jain, "Dam Vs. Drinking Water: Exploring the Narmada Judgement," Parisar, 2001.
25. Shiva, *Violence of the Green Revolution*.
26. Jain, "Dam Vs. Drinking Water."
27. Shiva, *Violence of the Green Revolution*.
28. "Punjab Floods Were Man-Made," *Economic Times* (Bombay), October 4, 1988.
29. Shiva, *Violence of the Green Revolution*, p. 145.
30. "Dams and Floods," *Indian Express*, October 21, 1988.
31. Shiva, *Violence of the Green Revolution*.
32. Marc Reisner, *Cadillac Desert: The American West and its Disappearing Water* (New York: Viking, 1986).

33. Vandana Shiva et al., *Ecology and the Politics of Survival* (New Delhi: Sage, 1991), pp. 202-240.
34. Ibid.
35. Government of India Agriculture Statistics, Delhi, 2000.
36. L. C. Jain, "Myths about Dams," (unpublished document, 2001).
37. Shiva et al., *Ecology and the Politics of Survival*, p. 186.
38. Ibid.
39. Arundhati Roy in "The Greater Common Good," *Frontline*, April 1999, p. 31.
40. Gandhian nonviolent, civil disobedience.
41. *Illustrated Weekly*, August 1984.
42. Vijai Paranjapaye, "Narmada Dams" (New Delhi: The Indian National Trust for Art and Cultural Heritage, 1987).
43. Ibid.
44. Ibid.
45. For an expanded discussion of the Ukai Dam and its social and ecological consequences, see Shiva, *Ecology and the Politics of Survival*, pp. 228-229.
46. Ibid.
47. For further discussion of the Bhakra Dam displacement, see Shiva, *Ecology and the Politics of Survival*.
48. Ibid., p. 230.
49. Maria Mies and Vandana Shiva, *Ecofeminism* (Halifax, NS: Fernwood Publications; London: Zed Books, 1993).
50. Ibid., p. 101.
51. *Dams and Development*, Report of the World Commission on Dams (London: Earthscan Publications, 2000), p. xviii.
52. Ibid.
53. Ibid., p. 18.
54. Letter from the anti-dam movement in Koel Karo.
55. On July 30, 2000, Kannada film star Rajkumar was kidnapped by the famous outlaw Veerappan. Veerappan presented 10 demands, including a mandate to find a permanent solution to the Kaveri water dispute. Other demands included making Tamil the second administrative language of Karnataka, unveiling the Thiruvalluvar statue in Bangalore, and increasing the daily wages for Manjolai Estate workers in Tirunelveli. Rajkumar was released on November 15, 2000.
56. Elizabeth Corell and Ashok Swain, "India: The Domestic and International Politics of Water Scarcity," in Leif Ohlsson, ed., *Hydropolitics; Conflicts over Water As a Development Constraint* (Dhaka: University Press; London: Zed Books, 1995), pp. 142-143.
57. Ibid, p. 143.
58. Ibid, p. 144.
59. Marq De Villiers, *Water: The Fate of Our Most Precious Resource* (New York: Houghton Mifflin, 2000), pp. 236-237.

60. Ibid., p. 239.
61. Michael Schultz in Ohlsson, ed., *Hydropolitics*, p. 106.
62. Ibid., p. 101.
63. Ibid., p. 99.
64. GAP is the Turkish Acronym.
65. Schultz in Ohlsson, ed., *Hydropolitics*, p. 99.
66. De Villiers, *Water*, p, 210.
67. Ibid.
68. Ibid, p. 11.
69. Schultz in Ohlsson, ed. *Hydropolitics*, p. 110.
70. Helena Lindholm, "Water and the Arab-Israeli Conflict," in Ohlsson, ed. *Hydropolitics*, p. 58.
71. Quoted in Saul Cohen, *The Geopolitics of Israel's Border Question*, (Boulder: Westview Press, 1986), p. 122.
72. Lindholm, "Water and the Arab-Israeli Conflict," p. 61.
73. Ibid, p. 69.
74. Ibid, p. 62.
75. Ibid, p. 63.
76. Ewan Anderson, "Water: The Next Strategic Resource," quoted in Lindholm, "Water and the Arab-Israeli Conflict," p. 77.
77. Fadia Darbes, Palestinian Water Authority, "Water Resources in the Region: An Approach to Conflict Resolution," (paper submitted to the P7 Summit on Water Issues, Brussels, June 7-10, 2000).
78. Military Order 158, November 19, 1967, Amendment to Water Law 31, 1953, quoted in Jerusalem Media Communication Center, 1993: p. 22.
79. Lindholm, "Water and the Arab-Israeli Conflict," p. 80.
80. *Mara Natha*, Secunderabad, India, March/April 2001.
81. De Villiers, *Water*.
82. Ibid, 216.
83. Ibid, p. 220.
84. Jan Hultin, "The Nile: Source of Life, Source of Conflict," in Ohlsson, ed. *Hydropolitics*, p. 29.
85. De Villiers, *Water*, p. 224.
86. Ibid, p. 225.
87. Imeru Tamrat, "Conflict or Cooperation in the Nile," (paper submitted to the P7 Summit on Water Issues, Brussels, June 7-10, 2000).
88. Ibid.
89. K. Tripathi, *Inter State River Conflicts* (Delhi: Law Institute, 1971), p. 31.
90. Hultin, "The Nile," p. 33.
91. The Helsinki rules were adopted by the International Law Association at the 52nd conference, held at Helsinki in August 1966. *Report of the Committee on the Uses of the Waters of International Rivers* (London: International Law Association, 1967).

92. *Report of Krishna Water Disputes Tribunal* (New Delhi: Government of India, 1973), p. 43.
93. Reisner, *Cadillac Desert.*
94. Shiva et al., *Ecology and Politics of Survival,* p. 255.
95. Ibid.

CHAPTER FOUR

The World Bank, the WTO, and Corporate Control Over Water

G iant water projects, in most cases, benefit the powerful and dispossess the weak. Even when such projects are publicly funded, their beneficiaries are mainly construction companies, industries, and commercial farmers. While privatization is generally couched in rhetoric about the disappearing role of the state, what we actually see is increased state intervention in water policy, subverting community control over water resources. Policies imposed by the World Bank, and trade liberalization rules crafted by the World Trade Organization (WTO), are creating a sweeping culture of corporate-states all over the world.

The World Bank: An Instrument for Corporate Control over Water

Not only has the World Bank played a major role in the creation of water scarcity and pollution, it is now transforming that scarcity into a market opportunity for water corporations. The World Bank currently has outstanding commitments of about $20 billion in water projects, $4.8 billion of which are for urban water and sanitation, $1.7 billion for rural water schemes, $5.4

billion for irrigation, $1.7 billion for hydropower, and $3 billion for water-related environmental projects.[1] South Asia receives 20 percent of World Bank water loans.

The Bank estimates the potential water market at $1 trillion.[2] After the collapse of the technology stocks, *Fortune* magazine identified the water business as the most profitable industry for investors.[3] Large corporations, such as the biotech giant Monsanto, covet this lucrative market. Monsanto is currently plotting its entry into the water business and is anxiously eyeing the funding available from development agencies:

> First we believe that discontinuities (either major policy changes of major trendline breaks in resource quality or quantity) are likely, particularly in the area of water, and we will be well positioned via these businesses to profit even more significantly when these discontinuities occur. Secondly, we are exploring the potential of non-conventional financing (non-governmental organisations, World Bank, USAID, etc.) that may lower our investment or provide local country business-building resources.[4]

The World Bank's use of loan conditions to privatize and trade water suits Monsanto well, and the two have already begun to talk of collaboration. Monsanto is "particularly enthusiastic about the potential of partnering with the International Finance Corporation (IFC) of the World Bank" and expects the IFC to "bring both investment capital and on-the-ground capabilities to our efforts."[5] For the company, sustainable development is the conversion of an ecological crisis into a market of scarce resources.

Monsanto estimates that the safe water market is worth billions of dollars. In 2000, the business of safe water provision was estimated to reach $300 million in India and Mexico. This is the amount currently spent by nongovernmental organizations (NGOs) for water development projects and local-government water-supply schemes; Monsanto hopes to tap these public finances for providing water to rural communities. Where the poor cannot pay, the company plans to create "non-tradition-al mechanism[s], targeted at building relationship[s] with local government and NGOs as well as through innovative financing mechanisms, such as microcredit."[6]

Monsanto also plans to penetrate the Indian market for safe water by establishing a joint venture with Eureka Forbes/TATA, a firm involved in water purification. The venture will help Monsanto control water delivery and distribution systems. The joint venture is ideal because it will allow Monsanto to "achieve management control over local operation[s] but not have legal consequences due to local issues"[7] Additionally, Monsanto is to purchase a Japanese company which has developed electrolysis technology for water treatment.[8]

In 1999, Monsanto ventured into aquaculture in Asia to aug-ment its agricultural biotechnology and expand its fish feeding and breeding capabilities. By 2008, the company expects to earn revenues of $1 billion and a net income of $266 million from its aquaculture business. While Monsanto's entry into aquaculture was justified under the auspices of sustainable development, industrial aquaculture is highly nonsustainable. The Supreme Court of India banned industrial shrimp farming because of its catastrophic consequences. Unfortunately, the government is trying to reverse the ban due to pressure from the aquaculture lobby. An Aquaculture Authority Bill has been presented to Par-liament to undo the environmental laws that protect the coast.[9]

Public-Private Partnerships: International Aid for Water Privatization

Privatization projects funded by the World Bank and other aid agencies are usually labeled "public-private partnerships." The label is powerful, both because of what it suggests and what it hides. It implies public participation, democracy, and account-ability. But it disguises the fact that public-private partnership arrangements usually entail public funds being available for the privatization of public goods.

Public-private partnerships can occur in the area of capacity building or management (operations and delivery of services). Management contracts can be short-term service contracts lasting six months to three years, longer contracts of three to five years with the public agency holding investment responsibility, or 25- to 30-year contracts with the private agency holding full operation, maintenance, delivery, and investment responsibility. Longer contracts usually involve bulk water-purchase agreements to be paid by the public agency, much like the power purchase agreements in energy privatization.

Public-private partnerships have mushroomed under the guise of attracting private capital and curbing public-sector employment. The World Bank, working on the assumption that the Third World will urbanize by 2025, estimates that $600 billion of investment in infrastructure projects will be required.[10] However, urbanization, like water privatization, is a possible result of World Bank policies, not an inevitable outcome.

Currently, public-private collaborations in water services receive millions of aid dollars. This money is a subsidy for private firms, who bid ferociously for the contracts. In India alone, there are 30 such collaborations in water services.[11] Public-private partnerships in the water business are meant to replace water services as a public service:

> First is the focus on *commercial orientation* through institutional reforms and restructuring. For example, a first step may be restructuring the water and sewage department on a profit center basis. Over time, corporatisation of the utility or separate joint venture companies to manage the water and sewage system will help to bring the necessary commercial orientation.
>
> The second aspect relates to the need for an appropriate regulatory framework. The basic objective of such institutional reform is to move towards a commercial and consumer orientation in service provision. The entire outlook changes from publicly provided free services as a right, to a consumer orientation with access to services.[12]

The erosion of water rights is now a global phenomenon. Since the early 1990s, ambitious, World Bank-driven privatization programs have emerged in Argentina, Chile, Mexico, Malaysia, and Nigeria. The Bank has also introduced privatization of water systems in India. In Chile, it has imposed a loan condition to guarantee a 33 percent profit margin to the French company Suez Lyonnaise des Eaux.[13]

Not only does privatization affect people's democratic right to water, it also affects the livelihoods and employment rights of those who work in municipalities and local water and sanitation systems. Public systems worldwide employ five to ten employees per 1,000 water connections, while private companies employ two to three employees per 1,000 water connections.[14] In most Indian cities, municipal employees have resisted privatization of water and sanitation services.

Privatization arguments have been based largely on the poor performance of public-sector utilities. Government employees are seen as excess staff, responsible for the low productivity of public water agencies.[15] The fact that poor public-sector performance is most often due to the utilities' lack of accountability is hardly taken into account. As it turns out, there is no indication that private companies are any more accountable. In fact, the opposite tends to be the case. While privatization does not have a track record of success, it does have a track record of risks and failures. Private companies most often violate operation standards and engage in price gouging without much consequence. In Argentina, two of the largest private French firms, Lyonnaise des Eaux and Compagnie Generale des Eaux, two of the largest private British firms, Thames Water and Northwest Water, and the largest public Spanish firm, Canal Isabel II, formed a consortium to bid for a World Bank-financed water privatization project. Employees at the public-sector utility provider Obras Sanitarias de la Nacion (OSN) in Buenos Aires were reduced from 7,600 to 4,000 in 1993. The unemployment of 3,600 workers has been touted as the most important achievement and indicator of success. While employment in water services went down, the price of water went up. Within the first year, water rates increased by 13.5 percent.[16]

In Chile, Suez Lyonnaise des Eaux insisted on a 35 percent profit.[17] In Casablanca, consumers saw the price of water increase threefold. In Britain, water and sewage bills increased 67 percent between 1989–90 and 1994–95. The rate at which people's services were disconnected rose by 177 percent. In New Zealand, citizens took to the streets to protest the commercialization of water. In South Africa, Johannesburg's water supply was overtaken by Suez Lyonnaise des Eaux. Water soon became unsafe, inaccessible, and unaffordable. Thousands of people were disconnected, and cholera infections became rampant.[18]

Despite its unpopularity among local residents worldwide, the rush to privatize water continues unabated. Encumbered by exorbitant debts, countries around the world are forced to privatize water. It is common for the World Bank and IMF to demand water deregulation as part of their lending conditions. Out of the 40 IMF loans disbursed through the International Finance Corporation in 2000, 12 had requirements for partial or full privatization of water supply and insisted on policy creation to stimulate "full cost recovery" and eliminate subsidies. In order to qualify for loans, African governments increasingly succumb to water privatization pressures. In Ghana, for instance, World Bank and IMF policies forcing the sale of water at market rate required the poor to spend up to 50 percent of their earnings on water purchases.[19]

The WTO and GATS: Trading Away Our Water

The General Agreement on Trade and Tariffs (GATT) was created along with the World Bank and IMF to manage the global economy in the postwar era. The 1944 Bretton Woods Conference gave shape to these institutions and instruments. GATT was intended to become the International Trade Organization in 1948, but the United States blocked the move since the rules of trade favored the South.[20] GATT therefore continued as an agreement until 1995, when the WTO was established on the basis of the agreements made at the Uruguay Round.

Before 1993, GATT dealt only with trade in goods beyond national borders. The Uruguay Round, negotiated between 1986 and 1993, expanded the scope of trade and the power of GATT by adding rules beyond goods and international trade. New rules were introduced on intellectual property, agriculture, and investment. Services were subjected to trade via the General Agreement on Trade in Services (GATS). By the time the WTO formed in 1995, the stage had been set for its unregulated power to override domestic policies and hijack common resources.

While the World Bank is promoting privatization of water through structural adjustment programs and conditions, the WTO is instituting water privatization via free-trade rules embodied in GATS. GATS promotes free-trade in services, including water, food, environment, health, education, research, communication, and transport. The WTO markets GATS as a "bottom up" treaty, citing the freedom of countries to liberalize trade progressively and to deregulate different sectors incrementally. In reality, GATS is a treaty with no reverence for or accountability to national democratic processes. In many cases, governments do not have the liberty to use cultural issues and resources in their WTO negotiations.

GATS not only bypasses government restrictions but also permits companies to sue countries whose domestic policy prevents free-market entry. For instance, in 1996 India passed the Provision of the Panchayats Act, recognizing the local community in tribal areas as the highest forms of authority in matters of culture, resources, and conflict resolution.[21] For the first time since India's independence, village communities (*gram sabhas*) were granted legal acknowledgment as community entities. Village communities retained a number of powers, including the power to approve or reject development plans and programs. *Gram sabhas* were also bestowed with the authority to grant land.

The act accepted the traditions of the people and their cultural identity by honoring their traditional relationship with the natural resources in their homeland. As the law stated, "a state legislation on the panchayats that may be made, shall be in consonance with the customary law, social and religious practices and traditional management practices of community resources."[22] The importance of having control over community resources was recognized not only as an economic necessity but as a touchstone of cultural identity: "Every gram sabha shall be competent to safeguard and preserve the traditions and customs of the people, their cultural identity, community resources and the customary mode of dispute resolution."[23]

The WTO disregards and even subverts hard-won victories such as the Indian Constitution. GATS is a tool to reverse the democratic decentralization to which diverse societies have been aspiring. GATS can challenge measures taken by central, regional, or local governments as well as nongovernmental bodies. Its rules are shaped entirely by corporations without any input from NGOs, local governments, or national governments.

The WTO and GATS: Facts and Fiction

On March 16, 2001, the WTO presented a defense of GATS in a press conference it called "GATS: Fact and Fiction." It argued that GATS does not violate rights to water, health, or education because it excludes "services supplied in the exercise of governmental authority." The WTO further maintained that GATS does not oblige countries to deregulate services or to open up their markets and that countries are free to tighten regulations on foreign investors.

A close examination of the WTO claims reveals a different and contrary reality. While GATS appears to exempt "services supplied in the exercise of government authority," it also mandates that such services be "supplied neither on a commercial basis, nor in competition with one or more service suppliers." Since "commercial basis" is not clearly defined, governments charging any tax or fees could be interpreted as engaging in commercial activity, and essential services could be dragged into a free trade ambit. Further, since most societies have pluralistic service providers, governments can be accused of being in competition with one or more service suppliers.

The "National Treatment" rule of GATS prohibits governments from discriminating between foreign and local service suppliers, even if the local provider is a community nonprofit and the foreign supplier is a giant water corporation. This rule also proscribes governments from requiring foreign corporations to hire or train citizens or to involve local people in management and ownership. Neither can these corporations be forced to transfer technology to local industries. The "Market Access" rule forbids governments to set limits on, among other things, the number of service suppliers, the value of service transactions or assets, the number of service operations, and the quantity of service output.

Water services have always been on the GATS agenda. For instance, "environmental services" currently include sewage, refuse disposal, sanitation, gas exhaust cleanup, and nature protection. At the heart of the environmental industry and of these services is of course water. The centrality of water to this field has been of interest not only to the WTO but also to the European Community—the government of the European Union. In 2000, the European Community reported that environmental services amount to $280 billion and are expected reach $640 billion by 2010, placing this sector in roughly the same category as the pharmaceutical and information technology industries.

The European Community has expanded the coverage of "water services" to include "water collection, purification and distribution."[24] And of course, as Ruth Caplan of the Alliance for Democracy points out, "collection could include the withdrawal of water from bodies of water and the extraction from ground- water and aquifers."[25] The proposals by the European Community could, therefore, have a major impact on community rights to water resources. At the Doha meeting of the WTO in November 2001, the United States sneaked water trade into the Ministerial Declaration. The section on Trade and Environment refers to "the reduction or, as appropriate, elimination of tariff and non-tariff barriers to environmental goods and services."[26] In other words, free trade in water.

New Agreements, Old Agenda

The WTO refers to GATS as the "first multilateral agreement on investment." Although, a global resistance defeated the Multilateral Agreement on Investment (MAI), the agenda has been resurrected by GATS. A similar free trade treaty is the North American Free Trade Agreement (NAFTA). Under NAFTA, Metalclad, an American waste management company, was able to extort $17 million from the Mexican government in a lawsuit. Metalclad's hazardous waste treatment and disposal site in the central Mexican state of San Luis Potos was shut down by local officials on the grounds that it was not environmentally sound. Unfortunately, NAFTA allows companies to sue governments for cash compensation if a country implements legislation that "expropriates" the company's future profits. Metalclad invoked this rule in its suit against the Mexican government and eventually won. The intense community opposition to Metalclad's facility was irrelevant.[27]

Corporate trade rights granted by trade agreements such as NAFTA and GATS apply to cases of corporate water ownership and control. NAFTA explicitly lists "waters, including natural or artificial waters and aerated waters" as tradable goods. And of course, as US trade representative Mickey Kantor pointed out in 1993, "when water is traded as a good, all provisions of the agreement governing trade in goods apply."[28]

In 1998, the American company Sun Belt Water sued the Canadian government for $10 billion because the company lost a contract to export water from Canada to California due to a 1991 ban on bulk water export imposed by the government of British Columbia.[29] The company claimed that British Columbia's ban on exports violated the protection of investor rights under NAFTA. The case is still under deliberation. Every level of government—including regional and local—is now forced to adhere to rules that it did not negotiate or agree to. Policy-making is no longer in the hands of local or national governments but in the grip of large multinational corporations. As Jack Lindsay, CEO of Sun Belt, puts it: "Because of NAFTA, we are now stakeholders in the national water policy in Canada."[30]

The Water Giants

Water has become big business for global corporations, which see limitless markets in growing water scarcity and demand. The two major players in the water industry are the French companies Vivendi Environment and Suez Lyonnaise des Eaux, whose empires extend to 120 countries. Vivendi is the water giant, with a turnover of $17.1 billion. Suez had a turnover of $5.1 billion in 1996.[31] Vivendi Environment is the "environmental services" arm of Vivendi Universal, a global media and communications conglomerate involved in television, film, publishing, music, the Internet, and telecommunication.

Vivendi Environment is engaged in water, waste management, energy, and transportation. In 2000, Vivendi Environment was awarded a 43-million-euro contract for a wastewater treatment plan in Berne, Switzerland. Vivendi also has a 50-50 joint venture company called CTSE in the Czech Republic. Total net sales are expected to be 200 million euros. Vivendi's subsidiary, Onyx, owns Waste Management Inc. Vivendi operates waste services in several countries, including Hong Kong and Brazil.

Other water giants include the Spanish company, Aguas de Barcelona, which dominates in Latin America, and the British companies Thames Water, Biwater, and United Utilities. Biwater was established in 1968 and given its name to reflect the company's involvement in both the dirty- and clean-water businesses. Thames is owned by RWE, an electric company whose ventures include water.

Biwater and Thames have operations in Asia, South Africa, and the Americas. In the 1940s, Biwater entered Mexico and the Philippines. By the 1970s, it had contracts in Indonesia, Hong Kong, Iraq, Kenya, and Malawi. By 1992, the Biwater empire had expanded to Malaysia, Germany, and Poland. In 2000, the company, along with a Dutch firm, launched its joint venture company, Cascal. Cascal has contracts in the United Kingdom, Chile, the Philippines, Kazakhstan, Mexico, and South Africa.[32] Another addition to the global water takeover is General Electric, which is working with the World Bank to create an investment fund to privatize power and water worldwide.

The privatization of water services is the first step toward the privatization of all aspects of water. The American water market for water supply and treatment, estimated at $90 billion, is the largest in the world, and Vivendi is investing heavily in order to dominate it. In March 1999, the company purchased US Filter Corporation for more than $6 billion and formed the largest water corporation in North America. Vivendi's projected revenue is $12 billion.[33]

Once the water giants enter the picture, water prices go up. In Subic Bay, the Philippines, Biwater increased water rates by 400 percent.[34] In France, customer fees increased 150 percent but water quality deteriorated; a French government report revealed that more than 5.2 million people received "bacterially unaccept-

able water."[35] In England, water rates increased by 450 percent
and company profits soared by 692 percent—CEO salaries increased by an astonishing 708 percent.[36] Service disconnection
increased by 50 percent.[37] Meanwhile, dysentery increased sixfold and the British Medical Association condemned water privatization for its health effects.[38]

In 1998, shortly after Sydney's water was overtaken by Suez
Lyonnaise des Eaux, it was contaminated with high levels of
giardia and cryptosporidium.[39] After water testing had been
privatized by A&L Labs, in Walkerton, Ontario, seven people, including a baby, died as a result of E. coli.[40] The company treated
the test results as "confidential intellectual property" and did
not make them public, just as Union Carbide withheld information about the leaked chemicals in its Bhopal, India, plant while
thousands were dying.[41] In Argentina, when a Suez Lyonnaise
des Eaux subsidiary purchased the state-run water company
Obras Sanitarias de la Nation, water rates doubled, but water
quality degenerated. The company was forced to leave the country when residents refused to pay their bills.[42]

The Great Thirst

In the maquiladoras of Mexico, drinking water is so scarce that
babies and children drink Coca-Cola and Pepsi.[43] Coca-Cola's
products sell in 195 countries, generating a revenue of $16 bil-
lion. Water scarcity is clearly a source of corporate profits. In an
annual report, Coca-Cola proclaims:

> All of us in the Coca-Cola family wake up each morning know-
> ing that every single one of the world's 5.6 billion people will get
> thirsty that day. If we make it impossible for these 5.6 billion peo-
> ple to escape Coca-Cola, then we assure our future success for
> many years to come. Doing anything else is not an option.[44]

Companies like Coca-Cola are fully aware that water is the
real thirst quencher and are jumping into the bottled water business. Coca-Cola has launched its international label Bon Aqua

(Dasani is the American version), and Pepsi has introduced Aquafina. In India, Coca-Cola's water-line is called Kinley. In addition to Coca-Cola and Pepsi, there are several other well-known brands such as Perrier, Evian, Naya, Poland Spring, Clearly Canadian, and Purely Alaskan.

In March 1999, in a study of 103 brands of bottled water, the Natural Resources Defense Council found that bottled water was no more safe than tap water.[45] A third of the brands contained arsenic and E. coli and a fourth merely bottled tap water. In India, a study conducted by the Ahmedabad-based Consumer Education and Research Center discovered that only three out the 13 known brands conformed to all bottling specifications.[46] None of the brands was free of bacteria, even though some claimed to be germ-free and 100 percent bacteria-free. Such false and misleading advertising has forced the Indian government to amend its Prevention of Food Adulteration rules to include bottled water. It now differentiates between mineral water obtained from and packaged close to a natural source and treated drinking water.[47]

The consequences of bottled water extend beyond price hikes and unsafe water. Environmental waste is a major cost incurred by the bottling industry. In the 1970s, 300 million gallons of bottled water were sold in nonrenewable plastic water containers. By 1998 this number had exceeded 6 billion. In India, the leading botded water producer Parle Bisleri accounts for 60 percent of the market share. It is expanding its $835,000 business and hopes to earn $208.8 million by 2002.

The head of Parle Bisleri, Ramesh Chauhan (also known as the "Water Baron"), has big plans: "Bisleri has to be made a megabrand. It's still *baccha* [a baby]. In the next two or three years, Bisleri must outsell both the cola companies together."[48] Chauhan forecasts that "the bottled water market will outstrip the carbonated drinks market in three years." Currently, bottled water accounts for 14 percent of the soft drink industry. Bisleri's one-liter bottle sells for 20 cents, and the five-liter bottle sells for

52 cents. Chauhan hopes to outsell Coke and Pepsi by keeping his prices lower.[49]

Bisleri, Pepsi, and Coke are not the only players in the Indian bottled water market. Britannia Industries and Nestle are also pushing their products, Perrier, San Pellegrino, and Price Life. Britannia markets Evian, which sells at $2 per liter, nearly double the hourly minimum wage. Evian is promoted as "an alternative beverage for lifestyle and fitness needs."[50] More than 500 rich families in India spend approximately $20 to $209 a month on Evian water. The Australian company Auswater Purification Ltd. is promoting its brand, Auswater. Smaller Indian companies like Trupthi, Ganga, Oasis, Dewdrops, Minscot, Florida, Aqua Cool, and Himalayan have also entered the market. These small firms account for 17 percent of the market share.

Global corporations are taking full advantage of the demand for clean water, a demand which has resulted from environmental pollution. Even though the corporations tap clean water resources in nonindustrialized, unpolluted regions, they refer to their bottling practice as "manufacture" of water. Nestle has a plant in Samalka in Haryana. In 1999, Pepsi started its Aquafina bottling plant in Roha, Maharashtra, and is setting up new plants in Kosi, Bazpur, Kolkata, and Bangalore. Coca-Cola bottles Kinley at its plants in Delhi, Mumbai, and Bangalore. The Indian packaged-water market is estimated at $104.4 million, with a growth of 50 to 70 percent per year.[51] In other words, bottled water production is expected to double every two years. Between 1992 and 2000, sales had increased from 95 million liters to 932 million liters.

As quickly as the water market is expanding in India, so is the traditional practice of giving water to the thirsty disappearing. For thousands of years, water was offered as a gift at *piyaos*, roadsides, temples, and marketplaces. Earthen pots known as *ghadas* and *surais* cooled the water during the summer for the thirsty, who would drink from their cupped hands. These pots have been replaced by plastic bottles, and the gift economy has been supplanted by the water market. No longer do all people have a right to quench their thirst; this is a right held exclusively by the rich. Even the president of India laments this misfortune: "The elite guzzle bottles of aerated drinks while the poor have to make do with a handful of muddied water."[52]

In Kerala, the restriction of water to the rich led local organizations to launch a campaign to boycott Coca-Cola. Partly as a protest and partly to develop alternative markets, residents of the coconut-rich state Kerala (*kera* means coconut in Malayalam) adopted the slogan "Goodbye Cola, Welcome Tender Coconut."[53] Coconut prices had dropped considerably when WTO rules flooded the region with soya and palm oil. Their low cost and their abundance made coconuts ideal for resisting another global conquest.

Corporations versus Citizens: Water Wars in Bolivia

Perhaps the most famous tale of corporate greed over water is the story of Cochabamba, Bolivia. In this semidesert region, water is scarce and precious. In 1999, the World Bank recommended privatization of Cochabamba's municipal water supply company, Servicio Municipal del Agua Potable y Alcantarillado (SEMAPA), through a concession to International Water, a subsidiary of Bechtel.[54] In October 1999, the Drinking Water and Sanitation Law was passed, ending government subsidies and allowing privatization.

In a city where the minimum wage is less than $100 a month, water bills reached $20 a month, nearly the cost of feeding a family of five for two weeks. In January 2000, a citizens' alliance called La Coordinadora de Defens a del Agua y de la Vida (The Coalition in Defense of Water and Life) was formed. The alliance shut down the city for four days through mass mobilization. Within a month, millions of Bolivians marched to Cochabamba, held a general strike, and stopped all transportation.[55] At the gathering, the protesters issued the Cochabamba Declaration, calling for the protection of universal water rights.[56]

The government promised to reverse the price hike but never did. In February 2000, La Coordinadora organized a peaceful march, demanding the repeal of the Drinking Water and Sanitation Law, the annulment of ordinances allowing privatization, the termination of the water contract, and the participation of citizens in drafting a water resource law. The citizens' demands, which drove a stake through the heart of corporate interests, were violently rejected. La Coordinora's fundamental critique was directed at the negation of water as a community property. Protesters used slogans like "Water Is God's Gift and Not Merchandise" and "Water Is Life."

In April 2000, the government tried to silence the water protests through martial law. Activists were arrested, protesters killed, and the media censored. Finally, on April 10, 2000, the people won. Aguas del Tunari and Bechtel left Bolivia, and the government was forced to revoke its hated water privatization legislation. The water company SEMAPA (along with its debts) was handed over to the workers and the people.[57] In the summer of 2000, La Coordinadora organized public hearings to establish democratic planning and management. The people have taken on the challenge to establish a water democracy, but the water dictators are trying their best to subvert the process. Bechtel is suing Bolivia, and the Bolivian government is harassing and threatening activists of La Coordinadora.[58]

By reclaiming water from corporations and the market, the citizens of Bolivia have illustrated that privatization is not inevitable and that corporate takeover of vital resources can be prevented by people's democratic will.

1. www.worldbank.org.
2. Maude Barlow, *Blue Gold: The Global Water Crisis and the Commodification of the World's Water Supply* (San Francisco: International Forum on Globalization, 2001), p. 15.
3. *Fortune Magazine*, May 2000.
4. Monsanto, "Sustainable Development Sector Strategy" (unpublished document, 1991), p. 3.
5. Ibid., p.14.
6. Monsanto, "Water Business Plan" (unpublished document, 1998).
7. Ibid.
8. Ibid.
9. Vandana Shiva, Afsar H. Jafri, and Gitanjali Bedi, *Ecological Costs of Economic Globalisation* (New Delhi: Research Foundation for Science, Technology, and Ecology, 1997), p. 45.
10. Riccardo Petrella, *The Water Manifesto: Arguments for a World Water Control* (London: Zed Books, 2001), p, 20.
11. Vandana Shiva et al., *License to Kill* (New Delhi: Research Foundation for Science, Technology, and Ecology, 2000), pp. 53-58.
12. Meera Mehta, *A Review of Public-Private Partnerships in the Water and Environmental Sanitation Sector in India* (New Delhi: Department for International Development, 1999), p. 7.
13. Barlow, *Blue Gold*, p. 15.
14. Emanuel Idelevitch and Klas Ringkeg, "Private Sector Participation in Water Supply and Sanitation in Latin America" (World Bank, 1995), p. 9.
15. Ibid.
16. Ibid, pp. 27-50.
17. Barlow, *Blue Gold*, p. 18.
18. Ibid.
19. Ghana National Coalition Against the Privatisation of Water, "Water is Not a Commodity," (unpublished document).
20. Ibid.
21. Provisions of the Panchayats (Extension to the Scheduled Areas) Act, 1996, Section 4(b).
22. Ibid, Sec. 4(a).
23. Ibid, Sec. 4(d).
24. GATS submission by European Union.
25. Ruth Caplan, "Alliance for Democracy" (paper circulated at the NGO GATS meeting, Geneva, April 2001).
26. WTO Doha Declaration (Ministerial Meeting, November 2000).
27. *New York Times*, July 31, 2000.
28. Ibid.
29. Ibid.
30. Quoted in Barlow, *Blue Gold*, p. 36.
31. Petrella, *The Water Manifesto*, p. 68.

32. Ibid.

33. Ibid.

34. Barlow, *Blue Gold*, p. 18.

35. Petrella, *The Water Manifesto*, p. 73.

36. Barlow, *Blue Gold*, p. 16.

37. Ibid.

38. World Development Movement (WDM), "Stop the GATSastrophe," November 2000, www.wdm.org.uk/cambriefs/wto/GATS.htm.

39. Barlow, *Blue Gold*, p. 17.

40. Ibid.

41. This information is based on my personal communication with Dr. Mira Shiva of the Bhopal Medical Relief Group.

42. Petrella, *The Water Manifesto*, p. 68.

43. Barlow, *Blue Gold*, p. 8.

44. "Small is Sustainable," International Society for Ecology and Culture, 2000, p. l.

45. Barlow, *Blue Gold*, p. 28.

46. Consumer Education Research Centre, *Insight* (January/February, 1998).

47. Government of India, PFA Amendment, 2000.

48. *Financial Express*, December 30, 2000.

49. *Business Times*, June 26, 2001, p. 10.

50. Ibid.

51. Ibid.

52. President Narayan's Republic Day speech, 1999.

53. I came across these slogans during a visit to Kerala.

54. Barlow, *Blue Gold*, p. 19.

55. Ibid.

56. See www.canadians.org/blueplanet/cochabamba-e.html.

57. Oscar Olivera and Marcela Olivera, "Reclaiming the Water" (unpublished document).

58. Ibid.

CHAPTER FIVE

Food and Water

Food and water are our most basic needs. Without water, food production is not possible. That is why drought and water scarcity translate into a decline of food production and an increase in hunger. Traditionally, food cultures evolved in response to the water possibilities surrounding them. Water-prudent crops emerged in water-scarce regions and water-demanding ones evolved in water-rich regions.

In the wet territories of Asia, rice cultures evolved and paddy field irrigation dominated. In the arid and semiarid tracts across the world, wheat, barley, corn, sorghum, and millet emerged as staples. In high-altitude regions, pseudocereals such as buckwheat provided nutrition. In the Ethiopian highlands, teff became the staple of choice. In deserts, pastoral cultivation was the basis of the food economy. Yet these diverse crops and agricultural styles are overlooked as food monoculture becomes the preferred method of production at the national, international, and corporate levels.

The water-use efficiency of crops is influenced by their genetic variation. Maize, sorghum, and millet convert water into biological matter most efficiently. Millet not only requires less water than rice, it is also drought-resistant, withstanding up to 75 percent soil moisture depletion. The roots of pulses and legumes allow efficient soil moisture utilization.

Since the Green Revolution, crops that produce higher nutrition per unit of water used have been called inferior and have been displaced by water-intensive crops. Water productivity has been ignored, the focus shifting to labor productivity. The replacement crops have produced not only unimpressive yields, but low organic matter, reducing the moisture conservation capacity of the soil.

Crop breeding in traditional societies took place, keeping in mind the effect of droughts. In a participatory breeding experiment with farmers in the desert region of Rajasthan, India, the International Center of Research in Crops for the Semiarid Tropics (ICRISAT) discovered that the farmers preferred their indigenous varieties of millet, citing the crop's resistance to drought. The farmers also chose their varieties because of higher biomass yield in the form of straw, manure, and animal feed. The modern industrial plant breeding had bred out the drought-resisting traits of crops.[1]

Industrial Agriculture and Water Crisis

Industrial agriculture has pushed food production to use methods by which the water retention of soil is reduced and the demand for water is increased. By failing to recognize water as a limiting factor in food production, industrial agriculture has promoted waste. The shift from organic fertilizers to chemical fertilizers and the substitution of water-prudent crops by water-thirsty ones have been recipes for water famines, desertification, waterlogging, and salinization.

Droughts can be aggravated by climate change and soil moisture reduction. Drought caused by climate change—a phenomenon known as a meteorological drought—is linked to rainfall failure.[2] But even with normal rain, food production can suffer if the soil moisture retention has been eroded. In arid areas, where forests and farms are entirely dependent on the recharge of soil moisture, addition of organic matter is the only solution.[3] Soil moisture drought occurs when organic matter necessary for moisture conservation is absent from soils. Prior to the Green Revolution, water conservation was an intrinsic part of indigenous agriculture. In the Deccan of South India, sorghum was intercropped with pulses and oilseeds to reduce evaporation. The Green Revolution replaced indigenous agriculture with monocultures, where dwarf varieties replaced tall ones, chemical fertilizers substituted organic ones, and irrigation displaced rainfed cropping. As a result, soils were deprived of vital organic material, and soil moisture droughts became recurrent.

In drought-prone regions, ecologically sound agricultural systems are the only way to produce sustainable food. Three acres of sorghum use as much water as one acre under rice paddy cultivation. Both rice and sorghum yield 4,500 kilograms of cereals. For the same amount of water, sorghum provides 4.5 times more protein, four times more minerals, 7.5 times more calcium, and 5.6 times more iron, and can yield three times more food than rice.[4] Had agricultural development taken water conservation into account, millet would not have been called a marginal or inferior crop.

The advent of the Green Revolution pushed Third World agriculture toward wheat and rice production. The new crops demanded more water than millet and consumed three times more water than the indigenous varieties of wheat and rice.[5] The introduction of wheat and rice has also had social and ecological costs. Their dramatic increase in water use has led to the instability of regional water balances. Massive irrigation projects and water-intensive farming, by adding more water to an ecosystem than its natural drainage system can accommodate, have led to waterlogging, salinization, and desertification. Waterlogging occurs when the water table falls 1.5 to 2.1 meters. If water is added to a basin faster than it can drain out, the water table rises. About 25 percent of the irrigated land in the United States suffers from salinization and waterlogging.[6] In India, 10 million hectares of canal-irrigated land is waterlogged and another 25 million hect-ares is under the threat of salinization.[7]

When waterlogging is recurrent, it is likely to lead to con-flict between farmers and the state. In the Krishna basin, water-logging at the Malaprabha irrigation project led to farmer rebellions. Before the introduction of the irrigation project, the semiarid land produced water-prudent crops such as joivar and pulses. The sudden climatic change, the intensive irrigation, and the cultivation of water-demanding cotton aggravated the prob-lem. Intensive irrigation of black cotton soils, whose water reten-tion capacity is very high, quickly created wastelands. While irrigation has been viewed as a means to improve land produc-tivity, in the Malaprabha area, it has had the opposite effect.[8] Farmers were shot by police when they refused to pay water taxes.[9] With the introduction of canal irrigation in the area, near-ly 2,364 hectares of land have become waterlogged and saline.

Salinization is closely related to waterlogging. The salt poi-soning of arable land has been an inevitable consequence of in-tensive irrigation in arid regions. Water-scarce locations contain large amounts of unleached soil;[10] pouring irrigation water into such soils brings the salts to the surface. When the water evap-orates, saline residue remains. Today more than one-third of the world's irrigated land is salt-polluted.[11] An estimated 70,000 hect-ares of land in Punjab are salt-affected and produce poor yields.[12]

The shift from rainfed food crops to irrigated cash crops like cotton was expected to improve the prosperity of farmers. Instead, it has led to debt.[13] Farmers borrowed money from banks for land development and for the purchase of seeds, chemical fertilizers, and pesticides. The total loans taken by the farmers increased from $104,449 in 1974 to more than $1.1 mil-lion by 1980. While farmers were struggling with unproductive land, banks were making payment demands. At the same time, irrigation authorities levied a development tax on water, known as a betterment levy. The latter increased from 38 cents to 63 cents per acre for *jowar,* and from 38 cents to over a dollar per acre for cotton. A fixed tax of 20 cents per acre was effective with or without water use.[14]

In March 1980, the farmers formed the Malaprabha Niravari Pradesh Ryota Samvya Samithi (Coordination Committee of Farmers of the Malaprabha Ittihsyrf Area) and launched a noncooperation movement to stop paying taxes.[15] In retaliation, government authorities refused to issue the certificates needed by the farmers' children to enroll in schools. On June 19, 1980, the farmers went on a hunger strike in front of a local official's office. By June 30, 10,000 farmers had gathered to support those on hunger strike. A week later, a massive rally was held in Navalgund, and farmers went on another hunger strike.

When no response came from the authorities, the farmers organized a blockade. About 6,000 farmers gathered in Navalgund, but their tractors were damaged and the rally was stoned by authorities. That same day, angry farmers seized the irrigation department, and set fire to a truck and 15 jeeps. The police opened fire, killing a young boy on the spot. In the town of Naragund, the police opened fire at a procession of 10,000 people, shooting one youth. The protesters responded by beating a police officer and a constable to death. The protests rapidly spread to Ghataprabha, Tungabhadra, and other parts of Karnataka. Thousands of farmers were arrested, and 40 were killed. In the end, the government ordered a moratorium on the collection of water taxes and the betterment levy.[16]

Unsustainable Agriculture: Water Waste and Destruction

The Aral Sea, the world's fourth-largest freshwater body, has been ruined by unsustainable agricultural activity. Rivers that recharge the lake are increasingly diverted toward the irrigation of 7.5 million hectares of cotton, fruit, vegetables, and rice.[17] Over the past few decades, two-thirds of the water has been drained away, salinity has gone up sixfold, and water levels have dropped by 20 meters. Between 1974 and 1986, the Syr Darya river never reached the Aral Sea; between 1974 and 1989, the Anu Darya failed to reach it five times. Instead, the water from these rivers feeds the Kara Kum irrigation canal near the Iranian border, 800 kilometers away.

In 1990, economist Vasily Selyunin commented of the Aral Sea: "The root of the problem is over irrigation, on a scale so vast that it has washed all the humus out of the soil. The loss had to be made good with shock doses of fertilizers. As a result, the earth has become like a junkie, unable to function without its fix." Fishing ports now lie 40 to 50 kilometers from the Aral shores, and the fish catch has collapsed from 25,000 tons a year to zero. Half of the population of the nearby city of Aralsk, Kazakhstan, has migrated. Unfortunately, as the Uzbek poet Muhammed Salikh points out, "You cannot fill the Aral with tears."[18]

Industrial farming is not just harming seas and rivers, it is impairing groundwater aquifers. The Ogallala Aquifer is irrigating farms in the High Plains of Texas. Each year, between 5 million and 8 million acre-feet of water are pumped from the Ogallala.[19] If the water continues to diminish at this rate, the only option left will be to shift to water-prudent, dryland farming or to abandon agriculture altogether. Sustainable agriculture policies would promote the former. Water markets promote the latter.

In the Third World, fossil fuel-based mining technologies have devastated water resources. Energized groundwater pumping promulgated by the Green Revolution was considered efficient in terms of energy and horsepower use. An irrigation pump powered by a 7.5 kilogram electric motor took five hours and one person to irrigate an acre of wheat; in contrast, a Persian wheel requires up to 60 bullock hours and 60 human hours.[20] Whether the water withdrawal was inconsistent with groundwater recharge was not given any weight in the calculations of efficiency. Energized pumps that desiccated large areas of prime farmland in less than two decades were seen as more effective than the traditional methods such as the Persian wheel, which had sustainably supported agriculture for centuries.

Many of the solutions proposed to the problem of agricultural water waste deny water for food production altogether. Industrial shrimp farming is a case in point. The most obvious and important impacts of industrial aquaculture are land and water salinization and drinking water depletion. Paddy fields once fertile and productive are turning into what local people call graveyards. This is true not just in India. In Bangladesh, where shrimp farming is widespread, the amount of rice production has dropped considerably. In 1976, the country produced 40,000 metric tons of rice; by 1986, production had plummeted to 36 metric tons.[21] Thai farmers report similar losses, harvesting 150 sacks of rice per year instead of the 300 sacks they were harvesting before the introduction of shrimp farms to the region.[22]

Women have been particularly affected by the proliferation of the shrimp industry. Land has become a scarce commodity, and fights over patches of land are more and more frequent. Women in Pudukuppam, India must walk one to two kilometers to fetch drinking water.[23] Wells have become sources of social tension. In the Indian village of Kuru, there is no drinking water available to the 600 residents due to salinization. After the 1994 protests by the local women, water was supplied in tankers, with each household receiving only two pots per day for drinking, washing, and cleaning. "Our men need 10 buckets of water to bathe after their fishing trips. What can we do with two pots?" is what women of coastal villages said to me that year. In Andhra Pradesh, the government supplied water by tankers from a distance of 20 kilometers for two years before it finally decided to move the 500 families to another location. In a number of regions, relocation was not possible and residents had no option but to use saline water for their crops and everyday needs.[24]

The United States is the most dramatic example of water waste in agriculture. In the western states, irrigation accounts for

90 percent of total water consumption. Irrigated land increased from 4 million acres in 1890 to nearly 60 million in 1977, of which 50 million were in the arid western states.[25] These areas are also affected by soil salinity because of salts dumped into rivers when irrigation waters drain. In a span of just 30 miles, the salt content of the Pecos River in New Mexico increases from 760 to 2,020 milligrams per liter.[26] In Texas, the salinity of the Rio Grande increases from 870 to 4,000 milligrams per liter in 75 miles.[27] Irrigation waters contribute 500,000 to 700,000 tons of salt annually to the Colorado River: the loss of yield due to salt is estimated at $113 million a year.[28] In San Joaquin Valley, California, crop yields have declined by 10 percent since 1970, an estimated loss of $312 million annually.[29]

Water exhaustion is not the only problem caused by industrial agriculture. In Bengal, India, deep tube-well drilling has been identified as the cause for arsenic poisoning. In west Bengal, more than 200,000 people are dying or are permanently maimed due to arsenic poisoning.[30] In Bangladesh, 70 million people are poisoned by arsenic; in 43 of Bangladesh's 64 districts, the arsenic level is around 0.05 milligram per liter and in 20 districts, the level is above 0.5 milligram per liter; the permissible limit is 0.01 milligram per liter.[31] Many villages report arsenic of up to 2 milligrams per liter, 200 times higher than the allowed level.

The Myth of Water Solution Through Genetically Modified Crops

In 2001, I attended the World Economic Forum (WEF) in Davos, Switzerland, where, at a session on water, a representative from Nestle suggested that genetic engineering would be a solution to water-intensive agriculture. He reasoned that genetic engineering could create drought-resistant crops that require little water. The obstacle, he argued, was the anti-genetic modification (GM) movement, which has prevented the introduction of drought-resistant varieties of GM crops.

The argument that genetic engineering will resolve the water crisis obscures two important points. First, peasants in drought-prone regions had bred thousands of drought-resistant crops, which were eventually displaced by the Green Revolution. Second, drought resistance is a complex, multigenetic trait, and genetic engineers have so far not been successful in engineering plants that possess it. In fact, the GM crops currently in the field or in labs will aggravate the water crisis in agriculture. For instance, Monsanto's herbicide-resistant crops, such as its Round-Up Ready soy beans or corn, have led to soil erosion. When all cover crops are killed by Monsanto's herbicide Round-Up, rows of soya and corn leave soils exposed to tropical sun and rain.

Similarly, the heavily advertised vitamin A–rich golden rice increases water abuse in agriculture. Golden rice contains 30 micrograms of vitamin A per 100 grams of rice. On the other hand, greens such as amaranth and coriander contain 500 times more vitamin A, while using a fraction of the water needed by golden rice. In terms of water use, genetically engineered rice is 1,500 times less efficient in providing children with vitamin A, a necessary vitamin for blindness prevention. The golden rice promise is what I call "a blind approach to blindness prevention."

The myth of water solution by way of GM crops obscures the hidden cost of the biotech industry—the denial of fundamental rights of food and water to the poor. Investing in indigenous breeding knowledge and protecting the rights of local communities are more equitable and sustainable ways to ensure access to water and food to all.

1. *Participatory Breeding of Millets* (The International Crops Research Institute for the Semi-Arid Tropics, 1995).
2. Vandana Shiva et al., *Ecology and the Politics of Survival. Conflicts Over Natural Resources in India* (New Delhi: Sage, 1991).
3. V. A. Kovda, *Land Aridization and Drought Control* (Boulder, CO: Westview Press, 1980); M. M. Peat and I. D. Teare, Crop-Water Relations (New York: Wiley, 1983).
4. Vandana Shiva, *Violence of the Green Revolution: Third World Agriculture, Ecology and Politics* (London: Zed Books, 1991), p. 70.
5. Shiva, *Violence of the Green Revolution*, p. 200.
6. Ibid.
7. Ibid.
8. Shiva et al., *Ecology and the Politics of Survival Conflicts Over Natural Resources in India* (New Delhi: Sage, 1991).
9. Ibid.
10. Unleached soils contain salts that are not washed away by rain.
11. Shiva, *Violence of the Green Revolution*, p. 128.
12. Ibid, p. 129.
13. Vandana Shiva et al., *Seeds of Suicide* (New Delhi: Research Foundation for Science, Technology, and Ecology, 2001).
14. Shiva, *Ecology and the Politics of Survival*, p. 234.
15. Ibid, 235.
16. Ibid.
17. Robin Clarke, Water: *The International Crisis* (Cambridge, MA: MIT Press, 1993), p. 61.
18. William Ellis, "A Soviet Sea Lies Dying," *National Geographic*, February 1990.
19. Marq De Villiers, *Water: The Fate of Our Most Precious Resource* (New York: Houghton Mifflin, 2000), p. 44.
20. Shiva, *Violence of the Green Revolution*, p. 141.
21. Vandana Shiva and Gurpreet Karir, *Chemmeenkettu* (New Delhi: Research Foundation for Science, Technology, and Ecology, 1997).
22. Ibid.
23. Ibid.
24. Ibid.
25. Tim Palmer, *Endangered Rivers and the Conservation Movement* (Berkeley, CA: University of California Press, 1986), p. 178.
26. Ibid, 192.
27. Mohamed T. El-Ashry, "Salinity Problems Related to Irrigated Agriculture in Arid Regions" (Proceedings of Third Conference on Egypt, Association of Egyptian-American Scholars, 1978), pp. 55-75.
28. El-Ashry, "Groundwater Salinity Problems Related to Irrigation in the Colorado River Basin and Ground Water" *Groundwater* Vol 18 No 1 January/February 1980, pp. 37-45.

29. De Villiers, *Water*, p. 143.

30. For further information on arsenic poisoning, visit the World Health Organization at www.who.int/water_sanitation_health/Arsenic/arsenic.htm.

31. For more reading on arsenic poisoning in Bangladesh, see Allan Smith, Elena Lingas, and Mahfuzar Rahman, "Contamination of Drinking-Water by Arsenic in Bangladesh: A Public Health Emergency," Bulletin of the World Health Organization, Vol. 78, No. 9 (2000), 1093-1103, available at www.who.int/bulletin/pdf/2000/issue9/bu0751 .pdf.

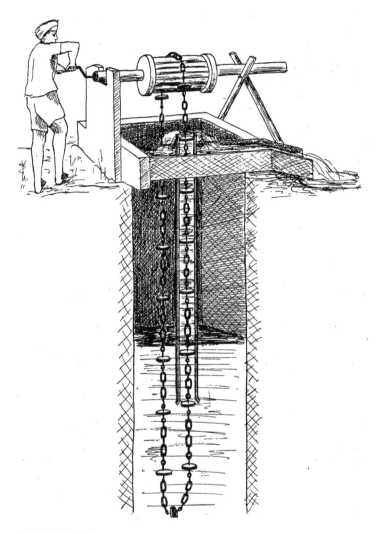

The two-bucket lift.

CHAPTER SIX

Converting Scarcity into Abundance

Scarcity and abundance are not nature given—they are products of water cultures. Cultures that waste water or destroy the fragile web of the water cycle create scarcity even under conditions of abundance. Those that save every drop can create abundance out of scarcity. Indigenous cultures and local communities have excelled in water conservation technologies. Today, ancient water technologies are once again gaining popularity.

Making the Desert Bloom

Like other desert regions, Rajasthan, the desert state of western India bordering Pakistan, has extremely low rainfall and very high temperatures. Unlike other desert regions, Rajasthan is blessed with water abundance. Anupam Mishra, the visionary behind the rejuvenation of the region's water system, observes:

> [I]f we compare the desertic region of Rajasthan to the deserts of the world, we notice that not only is it more populated but the very scent of life pervades it: In fact this region is considered as the most alive desert of the world.
>
> It is thanks to the local society that this is so. The people of Rajasthan did not mourn the lack of rain Nature bestowed

upon them. Instead they took it up as a challenge and decided to face it in such a way that from top to toe the people internalised the nature of water in its simplicity and its fluidity.[1]

Since every drop of rain has to be conserved, the indigenous knowledge is based on sensitive observation of rainfall and its patterns. The first drop of rain is called *hari*. Rain is also called *megaphusp* (cloud flower), *vristhi*, or *birkha*; water drops are called *bula* and *sikhar*. The *kuin, kuan, kundi, kund, tanka*, and *aagor* are diverse systems of water harvesting and water conservation, which make Rajasthan the most vibrant desert of the world. In this region, scarcity has been transformed into abundance through human ingenuity and labor. As Anupam Mishra notes, "Rajasthan's priceless drops of water are covered with sweat."

The culture of Rajasthan is not a culture of water deprivation but a culture of conservation, and "[n]owhere in the ancient history of Rajasthan can one find a description of its desert or even its other regions as dry, desolate or cursed land."[2]

Indigenous Water Management

There are more than 25 irrigation and drinking water systems built by the diverse communities of India. The *eri, keri, kunta, kulani, ahars, bandh, bandha, khadins, bundhies, sailata, kuthi, bandharas, low khongs, thodu, dongs, tanka, johad, nade, peta, kasht, paithu, bil, jheel*, and *talaks* are only a few of them. To this day, these ancient systems are the mainstay of survival in ecologically fragile zones.

The tank systems of southern India are some of the most enduring indigenous systems, lasting over centuries. They consist of several hundred linked reservoirs forming continuous chains that prevent water loss. These elaborate systems were impressive to the colonizers. Major Sankey, one of the first engineers of Mysore state, once remarked, "To such an extent has the principle of storage been followed that it would require some ingenuity to discover a site within this great area for a new tank."[3]

These tanks continue to play a central role in irrigation. In the Rayalseema region in the southern part of the Krishna basin, tanks irrigate 620,000 acres while major and minor irrigation projects cover 427,000 acres. In Anantapur, river water is diverted with the help of sand dams. Channels are also used for irrigation around India. In other regions, masonry dams called *panthams* are used for water storage.

Ahars and *pynes* are widely used for paddy-field irrigation in south Bihar. *Ahars* are built on drainage rivulets to collect water, and pynes arc used for capturing water from rivers running from the northern to the southern tip of the country. The effectiveness of these systems is notable. During the two great droughts of the late 1800s, Gaya district managed to survive because of its extensive *ahar* and *pyne* systems. The rest of Bihar, where these systems were not in use, was afflicted with famine.

In pre-British India, irrigation systems were managed by various social organizations within villages. Usually the membership of these organizations included the beneficiaries. In regions such as Maharashtra, irrigation systems were managed by water committees that maintained dams and de-silted canals. In Andhra Pradesh, the management systems known as *pinnapeddandarule* or *peddandarule* were run largely by youth, who provided hard physical labor. In Krishna district, where less labor-intensive work was involved, the membership rules were flexible, and de-silting, canal digging, and maintenance were equally shared by all the beneficiaries in proportion to the land they held. The committee fined those who failed to perform their share of work.[4]

Similarly, in south Bihar, both the construction and maintenance of water systems, known as *goam*, were collectively managed. The villagers were responsible for water allocation in their community. A system known as *parabandi* regulated the distribution of water among the villages from a common source. In cases involving large works, the rights of each village were formally recorded. In others, the regulations were largely customary and conflicts were resolved according to local procedures.

The British, whose agricultural system did not depend on irrigation, had no knowledge of water management when they arrived in India. Arthur Cotton, the founder of modern irrigation programs, even wrote:

> There are multitudes of old native works in various parts of India. These are noble works, and show both boldness and engineering layout. They have stood for hundreds of years. When I first arrived in India, the contempt with which the natives firstly spoke of us on account of this neglect of material improvements was very striking; they used to say we were a land of civilized savages, wonderfully expert about fighting, but so inferior to their great men, that we would not even keep in repair the works they have constructed, much less even imitate them in extending the system.[5]

Thomas Munro, who became governor of Madras in 1820, also acknowledged the extensive development of the indigenous water systems:

> To attempt the construction of new tanks is perhaps a more hopeless experiment than the repair of those which have been filled up through siltation. For there is scarcely a place where a tank can be made to advantage that has not been applied to this purpose by the inhabitants.[6]

The British, however, proceeded to control river water in India. In Rajasthan, they controlled water to maximize their salt revenues, to protect their transportation network, and to increase their agricultural income. In order to control rivers, the colonizers exerted force and dominance on those who depended on the river.

Decentralized Water Democracies

In 1957, the German historian and Marxian Karl Wittfogel published his famous *Oriental Despotism: A Comparative Study of Total Power*, in which he introduced the idea of a hydraulic society, a society where water management has historically been used to usurp power into a central entity.[7] The implication of Wittfogefs theory was that control over water implies control over people. Like his predecessor Karl Marx, Karl Wittfogel assumed that decentralized irrigation systems were linked to centralized power and that individuals conquering rivers became power elites. What Marx and Wittfogel failed to grasp was the freedom of cooperative management systems from dominant bureaucracies. That Indian irrigation systems relied on decentralized maintenance and not on centralized control was lost on these Western scholars.

Wittfogef's characterization of Asia's water systems has not gone unchallenged. Economic historian Nirmal Sengupta has pointed out that vast networks of irrigation systems are not necessarily large projects.[8] They can be a close-knit and locally managed network of microprojects. Sengupta has also shown that stagnation was not endemic to these traditional irrigation systems but that flexibility was central.[9] Cropping patterns changed annually according to water availability. With water resources under local control, decisions on land use were easier to make. Modern irrigation, on the other hand, uses centralized water control and distribution. Agricultural systems using modern dams are also less able to alter their cropping and irrigation practices to suit the availability of water. In addition, these large systems erode human rights and cause serious ecological damage.

Indifference to and ignorance about local ecological conditions led to the failure of many engineering projects during British rule. The Bradfield Dam catastrophe in Sheffield, England, in 1864 was a result of British expertise:

> A comparison naturally presents itself between the dam of the Bradfield reservoir, which failed, and the Indian model which has been so long and in so many instances successful, and, which if rightly constructed and faithfully attended to, may be regarded as ensuring the maximum of efficiency and safety.[10]

After 30 years of disastrous efforts to restore the Grand Anicut on the Kaveri River, Sir Arthur Cotton reverted to the more
effective indigenous methods. Cotton wrote:

> It was from [the Indians that] we learnt how to secure a founda-
> tion in loose sand of unmeasured depth. In fact, what we learnt
> from them made the difference between financial success and
> failure, for the Madras river irrigations executed by our engineers
> have been from the first the greatest financial successors of any
> engineering works in the world, solely because we learnt from
> them.... [W]ith this lesson about foundations, we build bridges,
> weirs, aqueducts and every kind of hydraulic work.... [W]e are
> thus deeply indebted to the native engineers.[11]

In traditional India, adequate and sustainable water supplies
were created under conditions of low and seasonal rainfall using
ancient ecological knowledge, technological expertise, and a culture of conservation. These sustainable water systems, however,
can be destroyed rapidly. Water technologies and water paradigms that fail to understand natural patterns can violate water
rhythms and degrade, deplete, and poison water resources.

People's Alternatives for Sustainability

While water privatization is the preferred policy by governments and global financial institutions, masses of people across
India and around the world are mobilizing to conserve water
and regain community control over their resources. The Pani
Panchayat movement, launched by the NGO Gram Gaurav Prat-
isthan (GGP), is an example of a people's movement that aims to
create an equitable and ecologically sustainable water system in
a drought-prone area.

The movement began in 1972, when Maharashtra was hit by
a severe drought. The lucrative and water-hungry cash crop sugarcane was diverting water away from people and nature. While
the government focused on famine relief and continued to rapidly exploit water resources, GGP founder Vikas Salunke recog-

nized the importance of strict water control and soil conservation as the most effective tools to survive the drought.

The Pani Panchayat believed in the rights of all residents to water. Water was considered a community resource, and the number of family members, not the size of one's land, determined how much water residents could receive. A suitable *patkari* (water distributor) was appointed to ensure fair day-to-day allocation. And while members of the Panchayat were free to decide how to use their water, sugarcane cultivation was regarded as an irresponsible use of resources and banned. A similar movement took root in 1982, when a group of migrant textile workers in Bombay returned to their villages to be greeted by drought, crop failure, and water shortage. Meanwhile, the government had plans to irrigate sugar plantations in 30 villages.

In response, the workers launched a movement called the Mukti Sangarsh, and mobilized more than 500 peasants to grow fodder for four months of the year on 2,000 acres of land and provide it free to the entire taluk, an administrative subdivision, if the government supplied the water. The villagers argued against the cultivation of water-intensive cash crops like sugarcane and advocated instead for equitable water distribution toward food-crop irrigation.

In 1985, 1,000 peasants participated in a march and pressed their demands. They also organized a conference on drought eradication that year. At the conference, the chairman of the Maharashtra State Drought Relief and Eradication Committee argued that if sugarcane cultivation were abandoned, 250,000 hectares of land could be irrigated, instead of the proposed 90,000 hectares. However, the sugar barons fiercely opposed the diversion of water away from cash-crop production. One politician's words reflect the sentiment of the sugar barons: "We will not give one drop of water from sugarcane; instead a canal of blood will flow. Cane and sugar factories are the glory of Maharashtra."[12]

After much resistance, the peasants gathered at Balawadi in 1989 to inaugurate the Baliraja Memorial Dam—a people's dam built with people's resources to meet people's needs. Popular participation prevented corruption, waste, and delay. The next step was to ensure the equitable distribution of water through social and collective control. Toward that end, the peasants agreed to stop sugarcane cultivation and instead plant mixed tree species on 30 percent of the land. They also opted to harvest staple grains using protective irrigation.[13]

In 1984, I visited the drought-stricken Maharashtra region. As a result of meager rainfall and devastated agriculture, people had resorted to brewing and selling illegal liquor for income. I learned that although the government spent $731.1 million on watershed development in Maharashtra, 17,000 villages had no water. I also discovered that the people's movement in Ralegaon Shindi had singlehandedly reversed desertification and economic collapse. Local residents had built water harvesting system made up of small dams, and they are now growing crops worth $146,000 to $188,000 a year. Illicit liquor sales have also tapered off.[14]

In the Alwar district of Rajasthan, water was being depleted at the rate of one meter a year, and the area was hit with a drought between 1985 and 1986. The youth organization Tarun Bharat Sangh mobilized people to rebuild *johads*, the traditional tank system for water harvesting. Local communities contributed $2.2 million and built 2,500 tanks in 500 villages. The water stored in a *johad* was to be shared by the entire village. The villages also decided how much land to irrigate and how much water to allocate to household use. The collective decision-making process over construction, maintenance, and use of water systems has helped prevent conflicts.[15]

Movements for water conservation are spreading all over India. In Gujarat, where nearly 13,000 villages have no dependable source of water and where groundwater is saline, women members of water councils are taking the lead in creating water harvesting systems. The people's investment in water conservation has also helped recharge groundwater, fill rivers, and increase crop production. In 1994, the Arvari River came back to life as result of recharge by 500 *johads*. Similarly, Ruparel, once a dead river, has been flowing since 1994 and is now the leading source of water for 250 villages. It was replenished by 250 *johads*.[16] In 2001, Tarun Bharat Sangh received the Magasasay Award for its work in water conservation.

The Swadhyaya movement of Gujarat, a movement aimed at self-development at all levels of organization, including individuals, communities, and countries, has led to the construction of 957 percolation tanks known as *nirmal neers*. As a result, close to 100,000 wells have been recharged. The Swadhyaya villagers endorse *bhakti*, the principle of volunteerism, and believe in 100 percent contribution. During the drought of 2000, Swadhyaya villages did not run out of water. Through their free labor and commitment to *bhakti*, the villagers have created an alternative to capital-intensive, nonlocal solutions to water scarcity.

Initiatives such as Swadhyaya, Tarun Bharat Sangh, Mukti Sangarsh, and Pani Panchayat indicate that water sustainability can emerge only from democratic control of water resources. Community control avoids ecological breakdown and prevents social conflict. Over the centuries, indigenous water management systems have relied on ancient knowledge and evolved into complex systems that ensure the equitable distribution of water.

Man-made water scarcity and ubiquitous water conflicts can be minimized with the recognition of water as a common resource. Water conservation movements are also showing that the real solution to the water crisis lies in people's energy, labor, time, care, and solidarity. The most effective alternative to water monopolies is water democracy. The current water war unleashed by multinational corporations can be won only through massive movements for water democracy. The blueprints provided by people's movements have shown the possibility of creating abundance out of scarcity.

1. Anupam Mishra, *The Radiant: Raindrops of Rajasthan*, translated by Maya Jani (New Delhi: Research Foundation for Science Ecology, 2001), p. 3.
2. Ibid.
3. S.T. Somasekhar Reddy, *Indigenous Tank System* (New Delhi: Research Foundation for Science, Technology, and Ecology, 1985).
4. Ibid.
5. Ibid.
6. Ibid.
7. K. A. Wittfogel, *Oriental Despotism; A Comparative Study of Total Power* (New Haven, CT: Yale University Press, 1957).
8. Nirmal Sengupta, *Managing Common Property: Irrigation in India and The Philippines* (New Delhi: Sage, 1991).
9. Ibid.
10. Quoted in Somasekhar Reddy, *Indigenous Tank System*.
11. Ibid.
12. Vandana Shiva, *Ecology and the Politics of Survival. Conflicts Over Natural Resources in India* (New Delhi: Sage, 1991).
13. Ibid.
14. Information is based on a personal conversation with Anna Hazare of Ralegaon Shindi, who had inspired a water revolution by mobilizing people.
15. Personal communication with Rajender Singh of Tarun Bharat Sangh, Alwar, May 2000.
16. Ibid.

CHAPTER SEVEN

The Sacred Waters

"Water is the source of all life."

—The Qur'an

"Apo hi stha mayobhuvas."
("Water is the greatest sustainer and hence is like a mother.")

—Taittiriya Samhita
The Sacred Ganges

Throughout history, water sources have been sacred, worthy of reverence and awe. The advent of water taps and water bottles has made us forget that before water flows through pipes and before it is sold to consumers in plastic, it is a gift from nature.

In India, every river is sacred. Rivers are seen as extensions and partial manifestations of divine gods. According to Rigvedic cosmology, the very possibility of life on earth is associated with the release of heavenly waters by Indra, the god of rain. Indra's enemy Vrtra, the demon of chaos, withheld and hoarded the waters and inhibited creation. When Indra defeated Vrtra, the heavenly waters rushed to earth, and life sprang forth.

According to Hindu mythology, the Ganges River originates in the heavens. The Kumbh Mela, a great festival centered around the Ganges, is a celebration of creation. According to one fable, the gods and demons were fighting over the *kumbh* (pitcher) filled with *amrit* (nectar), created by *sagar manthan* (the churning of the oceans). Indra's son Jayant ran away with the *kumbh*, and for 12 consecutive days, the demons fought the gods for the pitcher. Finally, the gods won, drank the *amrit*, and achieved immortality.

During the battle over the *kumbh*, five drops of *amrit* fell on earth at Allahabad, Haridwar, Nasik, and Ujjain, the four cities where the Kumbh Mela is still held. To this day, each city holds its own *mela* every 12 years. Allahabad's Maha Kumbh Mela in 2001 was one of the most spectacular festivals to date. Close to 30 million people gathered in the holy city to bathe in the sacred river Ganges.

The oldest and best-known myth about the creation of the Ganges is the story of Bhagirath. Bhagirath was the great-great-great-grandson of King Sagar, the ocean king. King Sagar had slain the demons on the earth and was staging an *aswameh-yagya* (a horse sacrifice) to declare his supremacy. Indra, the rain god and the supreme ruler of the kingdom of gods, feared losing his power of the *yagya* and stole Sagar's horse and tied it to the ashram of the great sage Kapil. At the time, Kapil was in deep meditation and unaware of Indra's mischief.

When King Sagar learned of his missing horse, he sent his 60,000 sons in search of it. The sons finally discovered the horse near the meditating sage and began to plot their attack on him. When the sage opened his eyes, he was angered to find the scheming brothers and reduced them to ashes.

King Sagar's grandson Anshuman was eventually successful in recovering the horse from Kapil. Anshuman reported to his grandfather that the sage had burned the 60,000 sons out of anger; the only way for the sons to reach their heavenly abode was if the Ganges could descend from Heaven so its water could cleanse the sons' ashes. Unfortunately, Anshuman and his son Dilip failed in bringing the Ganges to earth.

Finally, Anshuman's grandson Bhagirath went to the Himalayas and started meditating at Gangotri. After a long meditation, the Ganges appeared to him in bodily form and agreed to descend to the earth if someone could break her mighty fall, which would otherwise destroy the earth. King Bhagirath appealed to Shiva, who eventually agreed to use his hair to soften the descent of the Ganges. The river followed Bhagirath to where the ashes of King Sagar's sons were piled, purified their souls, and paved their way to the heavens.

Because the Ganges descends from heaven, she is a sacred bridge to the divine. The Ganges is a *tirtha*, a place for crossing over from one place to another. The *Gangastothra-sata-namavali* is an ode to the river, and reveals the profound effect of the river in India. The salute has 108 sacred names for the river.[1] The Ganges's role as mediator between this world and the divine is embodied in death rituals among Hindus. The ashes of our ancestors and kin are cast in the Ganges, so that, like the sons of Sagar, they too will be ensured a transition to the heavens. I was born and brought up in Doon Valley, bounded by the Ganges on the east and the Yamuna on the west. The rivers have nurtured me and shaped my sense of the sacred from childhood. One of the most moving experiences I had in recent years was immersing my father's ashes in the Ganges at Rishikesh.

Like the Ganges, the Yamuna, the Kaveri, the Narmada, and the Brahmaputra are all sacred rivers and are worshiped as goddesses. They are believed to cleanse and wash away spiritual and material impurities. Their reputed purifying characteristics are the reasons why, at their daily bath, devout Hindus chant, "O Holy Mother Ganga, O Yamuna, O Godavari, Oh Sarasvati, O Narmada, O Sindhu, O Kaveri. May you all be pleased to manifest in these waters with which I shall purify myself."

The Ganges does not merely possess the purifying qualities of water; it is saturated with antiseptic minerals that kill bacteria.

Modern bacteriological research has confirmed that cholera germs die in Ganges water. Dr. F. C. Harrison writes:

> A peculiar fact, which has never been satisfactorily explained, is the quick death, in three to five hours, of the *Cholera vibrio* in the waters of Ganga. When one remembers sewage by numerous corpses of natives, often cholera casualties, and by the bathing of thousands of natives, it seems remarkable that the belief of the Hindus, that the water of this river is pure and cannot be defiled and that they can safely drink it and bathe in it, should be confirmed by means of modern bacteriological research.[2]

It is no wonder that the Indian people hold the Ganges and other rivers dearly and believe they possesses mysterious powers. It is not surprising that despite the colonization of India by Coca-Cola and McDonald's, millions of people feel drawn to the Ganges on the occasion of Kumbh Mela.

An Ecological Tale

> Ganga, whose waves in swarga flow,
> Is daughter of the Lord of Snow.
> Win Shiva, that his aid be lent,
> To hold her in her mid-descent.
> For earth alone will never bear
> These torrents traveled from the upper air.[3]

The treks to the source of the Ganges are among my fondest memories of childhood. At an altitude of 10,500 feet stands the Gangotri, where a temple is dedicated to Mother Ganga, who is worshiped as both a sacred river and a goddess. A few steps from the Ganga temple is the Bhagirath Shila, a stone upon which King Bhagirath supposedly meditated to bring the Ganges to the earth. The shrine opens every year on Akshaya Tritiye, which falls during the last week of April or the first week of May. On this day, farmers prepare to plant their new seeds. The Ganga temple closes on the day of Deepavali, the festival of lights, and the shrine of the goddess Ganga is then taken to Haridwar, Prayag, and Varanasi.

The story of the descent of the Ganges is an ecological story. The above hymn is a tale of the hydrological problem associated with the descent of a mighty river like the Ganges. H. C. Reiger, the eminent Himalayan ecologist, described the material rationality of the hymn in the following words:

> In the scriptures a realisation is there that if all the waters which descend upon the mountain were to beat down upon the naked earth, then earth would never bear the torrents.... In Shiva's hair we have a very well known physical device, which breaks the force of the water coming down ... the vegetation of the mountains.[4]

The Ganges is not just a giver of peace after death—she is a source of prosperity in life. The Gangetic plain is one of the most fertile regions of the world. At the beginning of the ploughing season in Bihar, farmers, prior to planting their seeds, put Ganges water in a pot and set it aside in a special place in the field to ensure a good harvest. It is this treatment of the organic as sacred that inspired geographer Diana Eck to call the Ganges an "organic symbol." Eck writes:

> For the Ganga's significance as a symbol is not exhaustively narrative. First, she is a river that flows with waters of life in a vibrant universe. Narrative myths come and go in history.
>
> They may shape the cosmos and convey meaning for many generations, and then they may gradually lose their hold upon the imagination and may finally be forgotten. But the river remains, even when the stories are no longer repeated.[5]

Fourteen miles beyond Gangotri is Gaumukh, a glacier formed like the snout of a cow that gives rise to the Ganges. The Gaumukh glacier, which is 24 kilometers in length and six to eight kilometers in width, is receding at a rate of five meters per year. The receding glacier of the Ganges, the lifeline for millions of people in the Gangetic plain, has serious consequences for the future of India.

Christianity and the Sacred Waters

The sacredness of water has been inspired both by the power of rivers and by water as a life force. T. S. Eliot once wrote about the Mississippi River, "I do not know much about gods, but I think the river is a strong brown god."[6] All over the world we see the spiritual importance of water: in France, a temple sacred to the goddess Sequana is located at the source of River Seine, and the Marne River gains its name from Matrona, Divine Mother; the ancient name of the Thames River in England is Tamesa or Tamesis, denoting a river deity. In their book *Sacred Waters*, Janet and Cohn Bord list 200 ancient and holy wells in England, Wales, Scotland, and Ireland that have survived into modern times.[7]

Spiritual worship of water was wiped out in Europe with the rise of Christianity. The new religion called water worship pagan and denounced it as an abomination. At the Second Council of Arles, held around AD 452, a canon declared, "If in the territory of a bishop, infidels light torches or venerate trees, fountains or stones, and he neglects to abolish this usage, he must know that he is guilty of sacrilege."[8] In AD 960, Saxon King Edgar issued a decree requiring that "every priest industriously advance Christianity, and extinguish heathenism, and forbid the worship of fountains."[9] Such edicts continued to be issued well into the twelfth century.

In the 15th century, the Hereford Diocese Cathedral Registers passed a decree banning the worship of wells and other water sources in Turnaston, England:

> Although it is provided in the divine laws and sacred canons that all who shall adore a stone, spring or other creature of God, incur the charge of idolatry. It has come to our ears, we grieve to say, from the report of many credible witnesses and the common report of the people, that many of our subjects are in large number visiting a certain well and stone at Turnaston in our diocese where with genuflections and offer-

ings they, without authority of the Church, wrongfully worship the said stone and well, whereby committing idolatry; when the water fails they take away with them the mud of the same and treat and keep it as a relic to the grave peril of their souls and a pernicious example to others. Therefore we suspend the use of the said well and stone and under pain of greater excommunication forbid our people to visit the well and stone for purposes of worship. And we depute to each and all of you and firmly enjoin by virtue of holy obedience, to proclaim publicly in your churches and parishes that they are not to visit the place for such purposes.[10]

Despite the ban on water worship, people's deep faith in the sacredness of water persisted. In order to protect holy rituals, people converted sacred places for Christian use; old customs were absorbed into Christian rituals and water worship hid behind a Christian façade.[11] Water maintained its sacredness in rituals of baptism and hand washing. Baptism sites and churches were built close to and, at times, over wells.

Giving "Value" to Water

The word *value* is derived from the Latin term *valere* meaning "to be strong or worthy." In communities where water is sacred, the worth of water rests on its role and function as a life-force for animals, plants, and ecosystems. However, commodification of water reduces its value only to its commercial value. The *Oxford English Dictionary* now defines value in primarily economic terms: "that amount of some commodity, medium of exchange etc., which is considered to be an equivalent for something else; a fair or adequate equivalent or return." Like the term value, resources also has an interesting root. It originated from the word surge, meaning "that which has the capacity to rise again." Unfortunately, the term now means that which gains value as raw material for industry.

The proposal to give market values to all resources as a solution to the ecological crisis is like offering the disease as the cure. With the arrival of the industrial revolution, all value became synonymous with commercial value, and the spiritual, ecological, cultural, and social significance of resources was eroded. Forests were no longer living communities; they were reduced to timber mines. Minerals were no longer veins of the earth; they were merely raw material. We are now witnessing the commodification of two vital resources—biodiversity and water—which, for a long time, were beyond the reach of forest industrialization. Biodiversity is now a mere genetic mine and water a commodity.

The water crisis results from an erroneous equation of value with monetary price. However, resources can often have very high value while having no price. Sacred sites like sacred forests and rivers are examples of resources that have very high value but no price. Oceans, rivers, and other bodies of water have played important roles as metaphors for our relationship to the planet. Diverse cultures have different value systems through which the ethical, ecological, and economic behavior of society is guided and shaped. Similarly, the idea that life is sacred puts a high value on living systems and prevents their commodification.

Protection of vital resources cannot be ensured through market logic alone. It demands a recovery of the sacred and a recovery of the commons. And these recoveries are happening. A few years ago, a few thousand pilgrims used to walk from villages across north India to Hardwar and Gangotri to collect Ganges water for Shivratri, the birthday of the god Shiva. Carrying *kavads* (yokes from which two jars of holy water dangle and are never allowed to touch the ground) the *kavdias* now number in the millions. The highway from Delhi to my hometown, Dehra Dun, is shut during the weeks of the pilgrimage. Villages and towns put up free resting and eating places along the entire 200-kilometer pilgrimage route. The brightly decorated *kavads* containing Ganga water are a celebration of and dedication to the sacred.

No market economy could make millions walk hundreds of kilometers in the muggy heat of August to bring the blessings of the sacred waters to their villages. The 30 million devotees who went to bathe in the sacred Ganges for the Kumbh Mela did not see the value of the water in terms of its market price but in terms of its spiritual worth. States cannot force devotees to worship the water market.

Sacred waters carry us beyond the marketplace into a world charged with myths and stories, beliefs and devotion, culture and celebration. These are the worlds that enable us to save and share water, and convert scarcity into abundance. We are all Sagar's children, thirsting for waters that liberate and give us life—organically and spiritually. The struggle over the *kumbh*, between gods and demons, between those who protect and those who destroy, between those who nurture and those who exploit, is ongoing. Each of us has a role in shaping the creation story of the future. Each of us is responsible for the *kumbh*—the sacred water pot.

1. See appendix for a list of the 108 names for the Ganges.
2. Swami Sivananda, *Mother Ganges*, (Uttar Pradesh, India: The Divine Life Society, 1994), p. 16.
3. H. C. Reiger, "'Whose Himalaya? A Study in Geopiety," in T. Singh, ed., *Studies in Himalayan Ecology and Development Strategies* (New Delhi: English Book Store, 1980), p. 2.
4. Ibid.
5. Diana Eck, "Ganga the Goddess in Hindu Sacred Geography" in *The Divine Consort: Radha and the Goddesses of India*, John Stratton Hawley, Donna Marie Wulff, eds. (Berkeley: Graduate Theological Union, 1982), p. 182.
6. Uma Shankari and Esha Shah, *Water Management Traditions in India* (Madras, India: Patriotic People's Science and Technology Foundation, 1993), p. 25.
7. Janet Bord and Colin Bord, *Sacred Waters: Holy Wells and Water Lore in Britain and Ireland* (London; New York: Granada, 1985).
8. Ibid., p. 31.
9. Ibid.
10. Robert Mascall, *Bishop of Hereford*, pp. 1404-1417. Also see Bord and Bord, *Sacred Waters*, p. 45.
11. Ibid.

108 Names of the Ganges River

No.	Name	Meaning
1	Ganga	Ganges
2	Visnu-padabja-sambhuta	Born from the lotus-like foot of Vishnu
3	Hara-vallabha	Dear to Hara (Siva)
4	Himancalendra-tanaya	Daughter of the Lord of Himalaya
5	Giri-mandala-gamini	Flowing through the mountain country
6	Tarakarati-janani	Mother of the demon Taraka's enemy
7	Sagaratmaja-tarika	Liberator of the 60,000 sons of Sagara who had been burnt to ashes by the engry glance of sage kapila
8	Sarasvati-samayukta	Joined to the river Sarasvati (said to have flowed underground and joined the Ganges at allahabad]
9	Sughosa Melodius	Noisy
10	Sindhu-gamini	Flowing to the ocean
11	Bhagirathi	Pertaining to the saint Bhagiratha (whose prayers brought the Ganges down from heaven)

No.	Name	Meaning
12	Bhagyavati	Happy, fortunate
13	Bhagiratha-rathanuga	Following the chariot of Bhagiratha (who led the Ganges down to hell to purify the ashes of Sagara's Sons)
14	Trivikaram-padoddhuta	Falling from the foot of Vishnu
15	Triloka-patha-gamini	Flowing through the three worlds (i.e heaven, earth, and the atmosphere or lower regions)
16	Ksira-subhra	White as milk
17	Bahu-ksira	A cow which gives much milk
18	Ksira-vrksa-samakula	A boubding in the four "milk-trees," i.e., Naya-grodha (Banyan), Udumbara (glomerousfig-tree), and Madhuka (Bassia Latofolia)
19	Trilocana-jata-vasini	Dwelling in the matted locks of Siva
20	Trilocana-traya-vimocini	Releasing from the three debts, viz. 1. Brahma-carya (study of the Vedas) to th rishis 2. sacrifice and worship to the gods 3. procreation of a son, to the Manes
21	Tripurari-siras-cuda	The tuft on the head of the enemy of Tripura or Siva (Tripura was a triple fortification, built in the sky, air and earth of gold, silver and iron respectively, by Maya for the Asuras, and burnt by Siva)
22	Jahnavi	Pertaining to Jahnu who drank up the Ganges in a rage after it had flooded his sacrificial ground, but relented, and allowed it to flow from his ear
23	Nata-bhiti-hrt	Carrying away fear
24	Avyaya	Imperishable
25	Nayanananda-dayini	Imperishable

No.	Name	Meaning
26	Naga-putrika	Daughter of the mountain
27	Niranjana	Not painted with collyrium (i.e., color-less)
28	Nitya-suddha	Eternally pure
29	Nira-jala-pariskrta	Adorned with a net of water
30	Savitri	Stimulator
31	Salila-vasa	Dwelling in water
32	Sagarambusa-medhini	Swelling the waters of the ocean
33	Ramya	Delightful
34	Bindu-saras	River made of water-drops
35	Avyakta Unmanifest	Unevolved
36	Vrndaraka-samasrita	Resort of the eminent
37	Uma-sapatni	Having the same husband (i.e.,Siva) as Uma (Parvati)
38	Subhrangi	Having beautiful limbs (or body)
39	Srimati	Beautiful, auspicious, illustrious
40	Dhavalambara	Having a dazzling white garment
41	Akhandala-vana-vasa	Having Siva as a forest-dweller (hermit)
42	Khandendu-drta-sekhara	Having the crescent moon as a crest
43	Amrtakara-salila	Whose water is of nectar
44	Lila-lamghita-parvata	Leaping over mountains in sport
45	Virinci-kalasa-vasa	Dwelling in the water-pot of Brahma (or Vishnu or Siva)
46	Triveni Triple-braided	Consisting of the water of three rivers: Ganges, Yamuna, and Sarasvati
47	Trigunatmika	Possessing the three gunas

No.	Name	Meaning
48	Sangataghaugha-samani	Destroying the mass of sins of Sangata
49	Sankha-dundubhi-nisvana	Making a noise like a conch-shell and drum 50 Bhiti-hrt
50	Bhiti-hrt	Carrying away fear
51	Bhagya-janani	Creating Happiness
52	Bhinna-brahmanda-darpini	Taking pride in the broken egg of Brahma
53	Nandini	Happy
54	Sighra-ga	Swift-flowing
55	Siddha	Perfect, holy
56	Saranya	Yielding shelter, help or protection
57	Sasi-sekhara	Moon-crested
58	Sankari	Belonging to Sankara (Siva)
59	Saphari-puran	Full of fish (especially carp or Cyprinus Saphore, a kind of bright little fish that glistens when darting about in shallow water)
60	Bharga-murdha-krtalaya	Having Bharga's (Siva's) head as an abode
61	Bhava-priya	Dear to Bhava (Siva)
62	Satya-sandha-priya	Dear to the faithful
63	Hamsa-svarupini	Embodied in the forms of swans
64	Bhagiratha-suta	Daughter of Bhagiratha
65	Anatra	Eternal
66	Sarac-candra-nibhanana	Resembling the autumn moon
67	Om-kara-rupini	Having the appearance of sacred syllable, Om

No.	Name	Meaning
68	Atula	Peerless
69	Krida-kallola-karini	Sportively billowing
70	Svarga-sopana-sarani	Flowing like a staircase to Heaven
71	Sarva –deva-svarupini	Embodies about the continuance of peace
72	Ambhah-prada	Bestowing water
73	Duhkha-hantri	Destroying sorrow
74	Santi-santana-karini	Bringing about the continuance of peace
75	Daridrya-hantri	Destroyer of poverty
76	Siva-da	Bestowing happiness
77	Samsara-visa-nasini	Destroying the poison of illusion
78	Prayaga-nilaya	Having Prayaga (Allahabad) as an abode
79	Sita "Furrow"	Name of the eastern branch of the four mythical branches into which the heavenly Ganges is supposed to divide after falling on Mount Meru
80	Tapa-traya-vimocini	Releasing from the Three Afflictions
81	Saranagata-dinarta-paritrana	Protector of the sick and suffering who come to you for refuge
82	Sumukti-da	Giving complete spiritual emancipation
83	Siddhi-yoga-nisevita	Resorted to (for the acquisition of success or magic powers)
84	Papa-hantri	Destroyer of sin
85	Pavanangi	Having a pure body
86	Parabrahma-svarupini	Embodiment of the Supreme Spirit
87	Purna	Full

No.	Name	Meaning
88	Puratana	Ancient
89	Punya	Auspicious
90	Punya-da	Bestowing merit
91	Punya-vahini	Possessing or producing Merit
92	Pulomajarcita	Worshipped by Indrani, wife of Indra
93	Puta	Pure
94	Puta-tribhuvana	Purifier of the Three Worlds
95	Japa Muttering	Whispering
96	Jangama	Moving, alive
97	Jangamadhara	Support or substratum of what lives or moves
98	Jala-rupa	Consisting of water
99	Jagad-d-hita	Friend or benefactor of what lives or moves
100	Jahnu-putri	Daughter of Jahnu
101	Jagan-matr	Mother of what lives or moves
102	Jambu-dvipa-viharini	Roaming about or delighting in Rose-apple-tree Island (India)
103	Bhava-patni	Wife of Bhava (Siva)
104	Bhisma-matr	Mother of Bhisma
105	Siddha	Holy
106	Ramya	Delightful, beautiful
107	Uma-kara-kaamala-sanjata [Parvati]	Born from the lotus which created Uma
108	Ajnana-timira-bhanu	A light amid the darkness of ignorance

Index

About the Author

Photo by Kartikey Shiva

Vandana Shiva is a physicist, world-renowned environmental thinker and activist, and a tireless crusader for economic, food, and gender justice. She is the author and editor of many influential books, including *Making Peace with the Earth, Earth Democracy, Soil Not Oil, Stolen Harvest, Water Wars, and Globalization's New Wars*. Dr. Shiva is the recipient of more than twenty international awards, among them the Right Livelihood Award (1993); the John Lennon-Yoko Ono Grant for Peace (2008); The Sydney Peace Prize (2010); and the Calgary Peace Prize (Canada, 2011). In addition, she is a board member of the World Future Council and one of the leaders and board members of the International Forum on Globalization (whose other members include Jerry Mander, Edward Goldsmith, Ralph Nader, and Jeremy Rifkin). She travels frequently to speak at conferences around the world.